MODELS FOR MODALITIES

SYNTHESE LIBRARY

MONOGRAPHS ON EPISTEMOLOGY,

LOGIC, METHODOLOGY, PHILOSOPHY OF SCIENCE,

SOCIOLOGY OF SCIENCE AND OF KNOWLEDGE,

AND ON THE MATHEMATICAL METHODS OF

SOCIAL AND BEHAVIORAL SCIENCES

VOLUME 23

JAAKKO HINTIKKA

MODELS FOR MODALITIES

Selected Essays

D. REIDEL PUBLISHING COMPANY

DORDRECHT-HOLLAND / BOSTON-U.S.A.

Cloth edition: ISBN 90 277 0078 8
Paperback edition: ISBN 90 277 0598 4

Printed in The Netherlands by D. Reidel, Dordrecht

INTRODUCTION

The papers collected in this volume were written over a period of some eight or nine years, with some still earlier material incorporated in one of them. Publishing them under the same cover does not make a continuous book of them. The papers are thematically connected with each other, however, in a way which has led me to think that they can naturally be grouped together.

In any list of philosophically important concepts, those falling within the range of application of modal logic will rank high in interest. They include necessity, possibility, obligation, permission, knowledge, belief, perception, memory, hoping, and striving, to mention just a few of the more obvious ones. When a satisfactory semantics (in the sense of Tarski and Carnap) was first developed for modal logic, a fascinating new set of methods and ideas was thus made available for philosophical studies. The pioneers of this model theory of modality include prominently Stig Kanger and Saul Kripke. Several others were working in the same area independently and more or less concurrently. Some of the older papers in this collection, especially 'Quantification and Modality' and 'Modes of Modality', serve to clarify some of the main possibilities in the semantics of modal logics in general.

This work in the semantics of modality might at first sight seem to belie completely Quine's famous criticism of quantified modal logic. However, it soon leads to difficult and subtle choices between different approaches and assumptions. The problems one encounters here are very closely related to the true gist in Quine's suggestive but usually somewhat metaphorically expressed apprehensions. A diagnosis and a treatment of some of the main problems in this direction are outlined in 'Semantics for Propositional Attitudes' and especially in 'Existential Presuppositions and Uniqueness Presuppositions'. Although my suggestions, in particular my suggestion for the treatment of the difficulties, have incurred nothing like consensus among logicians, I am deeply convinced that they represent the most fruitful approach to this area for philosophical purposes. In

'Semantics for Propositional Attitudes' and elsewhere I try to give some glimpses of the wider philosophical implications of this approach.

In fact, philosophical uses of modal logics (especially of their semantics) and the philosophical methodology on which these uses are based constitute the focal interest of the present volume. My 1968 paper 'Epistemic Logic and the Methods of Philosophical Analysis', in which the typical structure of applications of logic to philosophy is discussed, has accordingly been included in the present volume as a kind of methodological preamble.[1] The same interest in the nature of philosophical applications of logical methods – applications often preached among contemporary philosophers but rarely practiced with any subtlety – is also in evidence in some of the other essays, particularly in the last essay, 'Deontic Logic and Its Philosophical Morals'. Its immediate predecessor in the present volume, 'On the Logic of Perception', offers an even more extended application of our semantical point of view to traditional philosophical problems. *Prima facie*, problems concerning sense data or the argument from illusion have little to do with modal logic. Hence I was myself struck how close – and how informative – an analogy can be found between the problems one faces in the semantics of modality and in the logic and epistemology of perception. Quine and Austin turn out to be allies in an almost identical fight against Church and Price, respectively, one is tempted to say.

On a more technical level, the uniqueness presuppositions which play (I shall argue) a crucial role in the semantics of modal logic turn out to be largely analogous with – or perhaps rather generalizations from – the existential presuppositions we encounter in the usual quantification theory. This connection is the main reason why a paper on 'Existential Presuppositions and Their Elimination' is also reproduced in this volume. Some of the philosophical insights we reach by making the existential presuppositions explicit are not much less interesting than those obtained in the model theory of modality. These insights are hopefully illustrated, though not exhausted, by the essay 'On the Logic of the Ontological Argument' included in the present volume.

A large number of small corrections and other changes have been made in the essays as they are printed here. One essay ('Modality and Quantification') has been considerably expanded. No complete uniformity of style, notation, or terminology has been attempted, although some of the most glaring discrepancies have been eliminated. An inveterate browser myself, I have sought to preserve the possibility of reading each essay separately. This is also the main reason why I have not tried to eliminate all overlap between the different articles. I hope, however, that this redundancy is not so great as to be seriously annoying. I am aware of being even more casual than usual with that fetish of second-rate logicians, quotes and use and mention. My appeal here is to the principle that one is to be considered innocent until one has been found guilty of an actual confusion caused by a failure to tell use from mention.

The bibliographical history of the several papers is indicated in an appendix, but the occasions on which I have presented them to various philosophical audiences in Scandinavia, the United States, England, Israel, and Austria, or otherwise discussed them with colleagues and students, are too numerous to be listed here. Collective thanks therefore have to replace most of the personalized ones. It must be said, however, that my main debts are probably to those Stanford students of different vintages whose questions, comments, and criticisms have forced me to clarify and develop further many of the theses presented in the present volume, and to Professor G. H. von Wright who many years ago first kindled my interest in modal logic.

Stanford, California, May 1969

JAAKKO HINTIKKA

[1] Although I can plead self-defence, some of the more polemical remarks in this essay may be slightly out of place in the present context. If so, I apologize to the philosophers in question.

TABLE OF CONTENTS

I. METHODOLOGICAL ORIENTATION

EPISTEMIC LOGIC AND THE METHODS OF PHILOSOPHICAL ANALYSIS

So-called ordinary language analysts and those philosophers who rely on the help of formal logic have often traded criticisms of each other's methods.[1] The store of specific examples and problems utilized in these exchanges seems to me remarkably small, however. An attempt to enrich our philosophical diet of examples might therefore be in order.

There is little disagreement as long as the applications of formal methods are restricted to the language of mathematics and of science. But when someone is bold enough to apply the techniques of symbolic logic to the analysis of such philosophically important concepts as necessity and possibility, knowledge, ignorance and belief, obligation and permission, etc., criticisms are likely to be directed not just against the details of one's analysis but against the very possibility of saying anything worthwhile about these concepts in formal terms.

As I have in effect observed elsewhere, criticisms of this kind have often a great deal of force *ad hominem* or perhaps rather *ad methodum*.[2] Far too many applications of the methods borrowed from logic have remained on the level of syntax (in Carnap's sense).[3] That is, their users have been content to put forward plausible-looking candidates for logical truth in terms of the concepts they are studying, and plausible-looking candidates for rules of deriving new logical truths from them. Plausibility here means agreement with whatever intuitions we happen to have concerning the concepts involved. The main trouble is that our intuitions, even when basically sound, frequently have to be applied in roundabout ways. These intuitions are often grounded, not on the logical relations to which they seem to pertain, but rather on certain more complicated logical relations. In another paper, I have demonstrated that certain proposed (and entirely intuitive) axioms for the logic of obligation lead to nonsensical results because the intuitions on which they were based pertained to the logical relations which would hold in a 'deontically perfect world' (i.e. a world in which all obligations are fulfilled) rather than to logical relations holding in the actual world.[4] The formalization of these (basically sound) intuitions therefore has to be accomplished by speaking of what ought to be the case, not of what is. In my book, *Knowledge and Belief*, I argued in the same way that many intuitions which we seem to possess about the inconsistency of statements expressed in terms of the notions of knowledge and belief are due, not to any real logical inconsistency of the statements they seem to pertain to, but rather to the logical impossibility of (consistently) believing or of knowing them.[5] Again, the formalization of these intuitions is a more complicated matter than first appears.

Such examples have convinced me that the usual straightforward axiomatization of the logic of philosophically interesting concepts is likely to be a worthless enterprise unless it is backed up by a deeper analysis of the situation. To obtain such an analysis, syntactical methods often have to be supplemented by semantical (model-theoretical) ones.[6] That is to say, we have to ask what conditions the truth of a set of statements imposes on the world, or (equivalently) what kinds of 'possible worlds' there must be in order for a set of statements to be consistent. Such a semantical analysis often gives us deeper insights into the logic of philosophically important notions. It seems to me that the critics of formal logic as a weapon of philosophical analysis have often overlooked the force of semantical methods and in effect spoken of syntactical methods only.

But even when semantical methods (or something equivalent) are employed, there is room for serious disagreement concerning the role of logical methods in philosophical analysis. I have been reminded of this fact by certain misunderstandings which have befallen my own attempt to sketch the logic of our central epistemic concepts by means of semantical techniques.[7] It seems to me that these misunderstandings can be traced to a view different from mine of the role of logical methods in philosophical analysis, and that a discussion of some of the problems which come up in this area might therefore serve as a good case study of the nature and applicability of these methods. At the same time I can outline the conception of the role of logical methods on which my book is based, although it unfortunately was not explained there explicitly.

The philosophically interesting concepts which we want to study are largely embedded in our ordinary usage. The question of the philosophical relevance of formal methods is thus closely related to the question of the applicability of these methods to the study of ordinary discourse.

What is the role of formal logic in this enterprise? It is often said, by philosophers as forceful and persuasive as Gilbert Ryle, that formal logic is a regimentation of the relevant sectors of ordinary discourse.[8] The aim of a branch of logic, say the logic of our epistemic concepts, is, according to these philosophers, to map as accurately as possible what we find in our ordinary talk about the same matters; in the case at hand, about what people know and believe. (Doubt already arises here. In my view, it has never been demonstrated satisfactorily that this is all that happens in such central areas of logic as quantification theory.) Often, this view of logic as the regimentation of certain features of our ordinary discourse is contrasted to the idea of logic as a revision of our ways with certain concepts, a map perhaps of what there is to be found in an ideal language rather than in an actual one.

Neither of these views seems to me to do justice to what is actually involved. A branch of logic, say epistemic logic, is best viewed as an *explanatory model* in terms of which certain aspects of the workings of our ordinary language can be understood. In some cases, this explanatory model may be thought of as bringing out the 'depth logic' which underlies the complex realities of our ordinary use of epistemic words ('knows', 'believes', etc.) and in terms of which these complexities can be accounted for. It therefore does not represent a proposal to modify ordinary language but rather an attempt to understand it more fully. But this explanatory model does not simply reproduce what there is to be found in ordinary discourse. As the case is with theoretical models in general, it does not seem to be derivable from any number of observations concerning ordinary language.[9] It has to be invented rather than discovered.

This conception of the relation of epistemic logic to ordinary language has many important consequences for the evaluation of what has been done in this branch of studies. At this point I shall mention only one example. Elsewhere, I have shown how one can make a general distinction between what is said of the individual who in fact is (say) *a* and what is said of *a*, whoever he is or may be, whenever one is speaking of propositional attitudes or using other modal concepts (in the broad sense of the term).[10] Now in ordinary discourse statements of the former type are usually more important and frequent than those of the latter type. Hence, if mere congruity with ordinary usage is what we want, it may seem advisable to restrict the substitution-values of our free individual symbols so as to allow statements of the former kind only. Suggestions in this direction have in fact been made for different reasons (e.g. by B. Rundle and by Dagfinn Føllesdal, in his unpublished dissertation), and there is something to be said for them. They can even be buttressed by applications of the Russellian theory of descriptions. This theory might seem (fallaciously) to enable us to eliminate constructions of the latter type altogether.[11] However, it seems to me that in this way we gain little insight into precisely why it is that we have to restrict the substitution-values of our free singular terms in the way we were asked to do. We also deprive ourselves of the possibility of characterizing the logical behavior of certain kinds of terms (pronouns, proper names) as distinguished from others. Hence in the interest of genuine theoretical insight into the logical situation we have to keep apart from ordinary language and carry out a deeper analysis of the situation.

Other more general aspects of the idea of logic as an explanatory model can also be registered. This idea is related to Wittgenstein's idea of a language-game. In fact, the explanations he offers in the *Blue Book* of the concept of a language-game appear to be in agreement with what I want to say of an explanatory model.[12] In many cases, an explanatory model may be thought of as giving us a way of using language in so far as this use is determined only by one main purpose which the part of language in question is calculated to serve. This shows why such an explanatory model does not accurately reflect what happens in ordinary discourse, for what happens there is also influenced by many other factors and pressures. Among them, there are factors of the following nature: (i) Other, competing purposes. Often, these are the vague general purposes which virtually all discourse is expected to serve or at least expected not to hinder. Thus many forms of discourse serve to keep others appraised of whatever the situation happens to be. This purpose is not served very effectively if the speaker does not make as full (explicit) statements as he is in the position to make.[13] (ii) Various pragmatic pressures, such as the pressure not to use circumlocutions without some specific purpose. (iii) Various built-in limitations of the human mind, e.g., the limitations of one's short-term memory. (iv) The pressures due to the particular context in which a sentence is uttered or written.[14]

The way in which one's explanatory model is supposed to throw light on what happens in ordinary discourse could be explained as follows: We shall call the meaning which an expression would have in the explanatory model its basic meaning. Now we may start from this basic meaning and see how it will be modified by the different factors (i)–(iv) (plus others, as the case may be). The resulting meanings, as far as they differ from the basic meaning, might be called residual meanings. If our explanatory model is an appropriate one, and if we have correctly diagnosed the pragmatic and the other extra factors involved in the different cases, we shall in this way be able to explain what actually happens on the different occasions of ordinary usage.

This is the strategy I followed in *Knowledge and Belief* when I tried to explain some of the different meanings which the locution 'knowing that one knows' may have in ordinary usage.[15] It was already pointed out there, however, that the problem we are dealing with is not restricted to epistemic logic.[16] Why are we justified in incorporating the law of double negation into our ordinary propositional logic? Surely in ordinary language a doubly negated expression very seldom, if ever, has the same logical powers as the original unnegated statement. Does not our propositional logic therefore distort grossly the logic of ordinary language? The answer is (very briefly) that if the basic meaning is assumed to be tantamount to that of the original unnegated expression we can explain the residual meanings which a doubly negated expression has on different occasions. Hence the basic meaning of a doubly negated expression can perfectly well be assumed to be the same as that of the original unnegated expression. No informed criticism of this point or of my explanation of the different residual meanings of 'knowing that one knows' seems to have been put forward.[17]

This way of trying to understand the workings of ordinary language may be contrasted to what is one of its most important rivals. This is the description of the meanings of our expressions in terms of paradigm cases.[18] Many philosophers of language who do not use the word 'paradigm' can also be classified as relying on essentially similar methods. Such a method is implicit in most appeals to 'what we normally say'. In spite of repeated farewells, it seems to me that this type of argument is still very much with us.

It is obvious that the paradigm-case method works in many cases.[19] However, it seems to me to be a very misleading approach to linguistic meaning in many others. A comparison with our method of explanatory models shows why this is the case. The basic meaning of an expression is not always, and perhaps not even usually, its normal (most frequent) meaning. It may even happen that an expression never has its basic meaning in ordinary language, at least not outside philosophers' discourse.

This seems to be the case very nearly with the expression 'knowing that one knows'. Several earlier philosophers have claimed that it is implied by (and therefore equivalent to) knowing *simpliciter*.[20] However, I doubt whether any such cases can be produced from ordinary discourse.

Those critics of *Knowledge and Belief* who have claimed that I there "conflate *knowing* with *knowing that one knows*" have therefore failed to understand the point I was arguing there.[21] Since they formulate their criticism in ordinary language terms, it has to be understood that they accuse me of conflating knowing and knowing that one knows in ordinary language. Far from conflating the two, however, I doubt very much whether they are *ever* equivalent in everyday discourse. What I did was to point out the reasons why the basic meanings of knowing and knowing that one knows are the same (at least in one important sense of knowing), and to argue that by understanding these reasons we can also understand some of the different ways in which their equivalence breaks down in ordinary language.

The case of 'knowing that one knows' illustrates another reason why a paradigmatic analysis of meanings is often doomed to be hopeless, if pursued seriously. There is no privileged residual meaning which could serve as the paradigm case. All the different residual meanings which I studied in *Knowledge and Belief* arise when the pragmatic pressures, e.g., the expectation that one does not use a circumlocution without some special purpose, lead to the breakdown of one or more of those features of the 'depth logic' of the expression 'knowing that one knows' which make it equivalent to knowing *simpliciter*. Since we know this logic, we can predict what these residual meanings are, for they go together with the different ways in which the equivalence can break down and which the depth logic brings out. None of these ways of breaking down seems to be privileged for theoretical reasons, and hence knowing one of them does not help one to understand the others.[22] For instance, how could one's familiarity with 'knowing that one knows' in the sense of 'being aware that one knows' possibly enable one to understand its use as meaning 'not merely being aware of a fact but also really knowing it'?

Furthermore, paradigmatic analysis is often much too inflexible to enable us to account for the meaning of a word in more complicated constructions. Professor A. R. White has accused me of misunderstanding the logic of the concept of awareness in *Knowledge and Belief*, without deigning to substantiate his charge in any way.[23] However, the explanations of the meaning of 'awareness' given in his own book *Attention* seem to me to fare much worse than anything that is said in *Knowledge and Belief*, if one tries to use them to explain the role of the term 'awareness' in slightly more complicated constructions.[24] Let us use the following passage of perfectly ordinary English as an example:

> Widmerpool could not have had the smallest notion of anything that had taken place between Jean Duport and myself, but people are aware of things like this within themselves without knowing their awareness. (Anthony Powell, *At Lady Molly's*, pp. 53–54, London 1957.)

About a third of what White says of awareness and being aware on pp. 42–43 of *Attention* is immediately thrown out of court by this single example. Does Widmerpool's diffuse awareness of whatever had taken place between Nicholas Jenkins (the narrator) and Jean Duport imply his being able to tell what it was? Surely not. Had the matter sometimes engaged Widmerpool's attention? Scarcely. Does attributing awareness to him suggest that the object of Widmerpool's awareness had frequently, or occasionally, come to his mind? The suggestion is rather to the contrary. Yet all these things are said in so many words by White about awareness.

In contrast to White's account of awareness, what I said in *Knowledge and Belief* already suffices to explain why our quotation from Anthony Powell (which I was not familiar with when I wrote my book) has the force it has. It is an instance of the denial of $(63)_2$ on p. 118 of *Knowledge and Belief*, combined with an assertion of the form '*a* is aware that *p*'. As explained in *Knowledge and Belief*, the joint force of the two is likely to be that *a* is aware that *p* but his awareness does not amount to real knowledge. By and large, this is also the thrust of what Powell lets Nicholas Jenkins say (or think). Widmerpool's awareness is simply too inarticulate, too heavily based on clues and half-conscious intimations to pass as genuine knowledge. In this respect, it is interesting to contrast our quotation from Powell with the one from Durrell which was given on p. 118, note 24, of my book, and which illustrated the denial of $(63)_1$, ('I knew but was not aware of knowing'). In Durrell's case, there is no indication that the grounds of the narrator were not always strong enough to support a claim to knowledge; the point is, rather, that they were not attended to by him. Here, Widmerpool's remarks show that the goings-on of Nicholas Jenkins had not remained a complete secret to him, yet his awareness does not amount to real knowledge. (There is probably also a suggestion in Powell that Widmerpool did not perhaps admit to himself his awareness.)

The notion of an expression's *having different senses* also turns out to be much more problematic than one first realizes.[25] The postulation of several irreducible senses of words and expressions is a typical device in an approach to the logic of ordinary language through the description of paradigm cases. In contrast, the differences in logical force that come about when a basic meaning is subjected to the pressure of various contextual and pragmatic factors are not unrelated different senses, and perhaps it would be wisest not to call them different senses at all. We have a clear-cut difference in sense when two variants of a concept have different logical properties in their basic use, i.e., different roles in our explanatory model. Then they will certainly have different basic meanings. But much of what normally passes as differences between different senses of words and expressions are simply differences between different residual meanings.

If a difference in sense means a difference in the 'depth logic', then in *Knowledge and Belief* there are only two basic senses of knowledge under discussion: in one of them (C.KK*) is satisfied; in the other it is not. A distinction of this kind cannot claim any novelty, however, for the difference in question is essentially the difference between knowledge and true belief, which philosophers have been discussing since Plato.[26] Of course, anyone interested in epistemic concepts has to study constructions in terms of knowledge and belief which have several residual meanings. The question is not so much whether we should call them different senses or not, but whether their logical behavior has to be postulated one by one or whether it is predictable by means of more general considerations. I have argued for the latter alternative.

Yet I have been accused of postulating all kinds of different senses of knowing.[27] This may be due to the fact that I listed in *Knowledge and Belief* several different pronouncements of various philosophers on the concept of knowledge; but the purpose of these references was not to make distinctions, but rather to argue that any strong concept of knowledge has to satisfy (C.KK*), however it is characterized.[28] Hence I cannot see anything but misunderstanding in these charges, which in fact sometimes strikingly illustrate the superiority of my approach to others. For instance, Mr. Max Deutscher criticizes me for trying to set up different "senses" of the expression, '*a* does not know that *p*'. The right thing, Deutscher avers, is to say simply that this expression sometimes carries the suggestion that not-*p* is in fact the case. This statement brings out strikingly how limited the purposes are which Deutscher reckons with and which he also tacitly (but wrongly) imputes to me. What he says is obviously and completely useless as a seriously intended explanation of what goes on in ordinary discourse.[29] To try to 'explain' the different logical force which a statement of the form '*a* does not know that *p*' has on different occasions by saying that it sometimes carries a "suggestion" that *p* is the case would be completely on a par with an explanation of the sleep-inducing properties of opium in terms of its 'dormitive' virtues – if it were correct. As it is, however, Deutscher fares worse than Molière's learned doctor in that there sometimes is much more than a mere "suggestion" of the truth of *p* involved. If I ask: "Why is John rushing to the airport?" and draw the retort: "He does not know that the SAS pilots are on strike", then this reply does not merely "suggest" that a strike might be on; it presupposes that it is.

In my book I sketched an explanation of the difference between those occasions on which '*a* does not know that *p*' implies not-*p* and those on which it does not.[30] This explanation was formulated in terms of the ancient role of 'that' as a demonstrative. It therefore turned on the context in which the statement was made, for it is this context that supplies the object of reference for the demonstratively construed 'that'. This explanation has nothing to do with different basic meanings of any expression. Whether or not it is correct, it strives to *explain* certain plain differences in the logical behavior of one and the same expression on different occasions, and not to pretend to have accomplished something by introducing a merely verbal distinction between different cases. In fact, it is Deutscher who is forced to introduce different senses here, for what else can there be to distinguish those cases in which the 'suggestion' is present from those in which it is not than a difference in sense? Instead of criticizing me, he thus succeeds merely in illustrating strikingly the vacuousness of the alternative he offers to my approach as a vehicle of serious theorizing about ordinary language.

In *Knowledge and Belief* I did not suggest any terminological distinction between the different senses of 'different senses'. If a distinction is called for, it might be appropriate to refer to the different basic senses of a word or of an expression as different senses and to different residual meanings as different uses. In other words, a difference between the different logical powers a word or an expression has on different occasions would indicate a real difference in sense if and only if it cannot be explained away (in terms of such factors as (i)–(iv)) but has to be reproduced in our explanatory model. A difference which can be accounted for without postulating more than one basic meaning would be a mere difference in use. We could thus say, e.g., that a double negation does not have different senses on different occasions, but that it is used in many different ways. Grammars in fact mention some of them, e.g., the use of a double negative to indicate hesitation or uncertainty. This is not the only current use of a double negative, however. A fairly small sample easily yields other uses, such as signalling diffidence (which is not identical with uncertainty!) or expressing irony. No one of these residual meanings (different uses) helps to understand the others, which makes a paradigmatic analysis of the meaning of a double negative completely useless.

In fact, what first gives the appearance of several unrelated senses of a word or an expression is often a symptom of the presence of nothing more than different uses (different residual meanings), accountable for in terms of one basic meaning. For instance, the variety of different things that 'knowing that one knows' can serve to express already strongly suggests that we have to do with a mere difference in use and not with genuinely different senses. It is interesting to see that a majority of those philosophers who have considered the matter have opted for precisely the same basic meaning (knowing that one knows identical with knowing *simpliciter*) as was tentatively proposed in my book.

What is the basis of this basic meaning? Formally, it turns on the condition (C.KK*) adopted in my book. However, the force of this condition is essentially just that of making sure that knowing implies knowing that one knows. Hence the mere adoption of this condition does not illuminate things in this respect. In *Knowledge and Belief* I offered a few considerations suggesting that this condition has to be adopted when a strong sense of knowing is used.[31] What has been said in the present paper gives us a starting point for another line of argument for the same conclusion.

The explanatory model which an epistemic logic strives to be was said to be a model of the use of our epistemic concepts, such as it would be if it were governed merely by the basic purposes for which we have these concepts in the first place. But what is the purpose of having a strong sense of knowledge in our language? What is this notion good for? The best answer to this question seems to have been given by Douglas Arner.[32] He may be said to be pushing further a line of thought suggested by James Urmson in his interesting paper, 'Parenthetical Verbs'.[33] Urmson argues there that the function of such 'parenthetical verbs' as 'knows' and 'believes' is to indicate the evidential situation in which a statement is made. (Here I am not concerned with the alleged parentheticity of Urmson's parenthetical verbs.[34]) What Arner does is just to try to state more explicitly what this evidential situation has to be in the case of the verb 'knows'. He argues that for someone to know that *p* his evidence (or his grounds – the term is not at issue here) has (have) not only to be good but as good as it (they) can be. It has to be such that further inquiry loses its point (in fact, although it is logically possible that such an inquiry might make a difference). The concept of knowledge is in this sense a 'discussion-stopper'. It stops the further questions that otherwise could have been raised without contradicting the speaker.

One such question is the following, directed to whoever is putting forward a knowledge-claim: 'You say that your grounds are conclusive. I accept them. But do you have conclusive grounds for saying that the grounds you have are in fact conclusive? Is some further inquiry perhaps needed in order to assure this?' This is a perfectly legitimate, albeit apparently somewhat recondite type of challenge. It will not even be recondite if what is being claimed is in so many words 'real' knowledge as distinguished from true opinion. If the notion of knowledge is to serve as a definite discussion-stopper, challenges of this sort must be precluded. The critical condition (C.KK*) formulates precisely that feature of the logic of knowledge which stops them. Without it, the concept of knowledge would not serve fully the basic purpose it is (in the language-game so ably described by Arner) calculated to serve, and which is clearly one of the (many) purposes it in fact serves in ordinary discourse.

The condition (C.KK*) also serves to bring out part of what is true in the idea that knowledge – 'real' knowledge – presupposes *certainty*. The objective element in the notion of certainty seems to be that further investigation has lost its point, and what (C.KK*) does is just to rule out certain kinds of further inquiry.

Those who assume that knowing implies knowing that one knows have sometimes been accused of generating an infinite regress (or perhaps better an infinite progress of levels or orders of knowledge). They seem to be postulating an infinity of separate acts of knowing taking place simultaneously. The opposite, however, is what really is going on. To say that knowing (in its basic sense) logically implies knowing that one knows, is to say that a claim to knowing that one knows does not add anything at all to an ordinary knowledge-claim, but merely makes it in a roundabout fashion (and therefore invites the ascription of some residual meaning to what one says). Thus the condition (C.KK*) has precisely the effect of making an infinite series of higher-and-higher orders of knowledge impossible in principle and not just in practice.

The argument which I just gave for (C.KK*) is closely related to the arguments which Professor Sören Halldén has given for the analogue of this assumption for logical modalities (necessity implies necessary necessity).[35] His interesting and able argument is based on the requirements which the concept of necessity has to fulfil if it is to serve the purposes for which we have it in our language in the first place. The general mode of his argument is therefore similar to mine, although he does not consider the possible discrepancies between what the primary function of a concept presupposes and what we actually find in ordinary language. It seems to me, however, that Halldén ought to have formulated his arguments in terms of knowledge and not in terms of logical necessity. The concept of logical necessity, it seems to me, is largely determined by the semantics of our ordinary, non-modal logic, and is therefore not amenable to pragmatic requirements of the kind Halldén considers. Nevertheless his arguments would apply to the concept of knowledge virtually intact.[36]

By way of conclusion, it may be pointed out that my term 'explanatory model' is perhaps a shade too humble. Explanatory models of the sort I have discussed are in effect essentially what in many other walks of life – or at least of scholarship – are known as theories. It seems to me that the difference between an approach to the logic of ordinary language in terms of my 'explanatory models' and an approach to it in terms of 'what we ordinarily say' or in other paradigmatic terms is to a large extent a difference between a genuine theory of the meaning of the words and expressions involved and a mere description of the raw data of the language. In fact, the objections to the use of formal logic in the analysis of ordinary language concepts are in my view often merely special cases of the confusion of those who think that a theory is nothing but a summary of the data which we have in some area of investigation.

Of course, the actual success of what I have called 'explanatory models' will depend on how much interesting structure there is hidden in our ordinary language and in our ordinary ways of employing it, structures sharp and general enough to be amenable to semantical treatment. In this respect, I am much more optimistic than many other philosophers of language, although I am willing to confess that this optimism is not backed up by as many concrete results as I should like to see. Paradoxically, it is very often ordinary language philosophers who are pessimistic concerning the possibility of uncovering interesting structures underlying the complex realities of ordinary discourse. Because of this pessimism, they conceive of their work as merely reporting the obvious data of use and usage. Personally, and as a matter of research strategy, this strikes me as a counsel of the bleakest philosophical despair, as a denial of all theoretical interest of the phenomena they are dealing with. The trouble with many ordinary language philosophers seems to be that they neither take ordinary language seriously as an object of theoretical study nor trust it enough to try to go beyond its surface so as to find interesting, generalizable structures.

In this paper I have tried to assemble reminders to show that this despair is perhaps unjustified.

REFERENCES

[1] This statement is admittedly oversimplified. In fact, many logicians have disclaimed all direct applicability of their considerations to ordinary usage, and by so doing avoided all confrontation with the ordinary language analysts, some of whom in fact are not in principle averse to the use of logical methods. Some of the sharpest criticism of the methods of so-called ordinary language philosophers comes from scholars whose main allegiance is to some other set of technical methods (e.g., to empirical semantics or to structural linguistics) or who are not practising logicians in spite of their recognition of the philosophical relevance of formal logic.

For reasons to be indicated later, much of the criticism of the methodology of ordinary language philosophers has been focused on what is known as the paradigm-case argument. Glimpses of this discussion are seen from Antony Flew, 'Again the Paradigm', in *Mind, Matter and Method* (ed. by Paul K. Feyerabend and Grover Maxwell), University of Minnesota Press, Minneapolis, Minn., 1966, pp. 261–272, which contains a short bibliography of the subject.

[2] See e.g. the early paragraphs of the following papers: 'The Modes of Modality', *Acta Philosophica Fennica* **16** (1963) 65–82; 'On the Logic of Existence and Necessity', *The Monist* **50** (1966) 55–76. These papers are reprinted in the present volume, pp. 71–86 and 23–44, respectively.

[3] A great deal of confusion has been created by oversimplified applications of the Carnapian trichotomy syntax-semantics-pragmatics in other respects, too. Such oversimplified uses have fostered the illusion that all study of the uses of language must lie beyond the purview of logical methods and belong to the psychology or sociology of language rather than to logic or philosophy. There is not a shred of a reason, however, why the general structures exhibited by language *in use* could not also be studied by logical and mathematical means.

Following scattered clues found in the writings of some of the most eminent logicians, I have suggested elsewhere that there is an intimate connection between the central and important quantification theory of modern logic and certain types of games in the precise sense of (mathematical) game theory. In the same spirit, I shall suggest later in this paper that formal logical structures are often apt to illuminate the workings of ordinary discourse in so far as they can be conceived of as exhibiting the structure of certain possible ways of using language – using it not just for its own sake but to some purpose.

Although Carnap is not to be blamed for the unfortunate confusion his trichotomy has generated, it seems to me that it is time to replace the trichotomy by some more flexible scheme.

[4] See the last parts of my paper, 'Quantifiers in Deontic Logic', *Societas Scientiarum Fennica, Commentationes Humanarum Litterarum* **23** (1957), no. 4. The same point is made, and elaborated, in the new paper 'Deontic Logic and Its Philosophical Morals' in the present volume, pp. 184–214.

[5] *Knowledge and Belief: An Introduction into the Logic of the Two Notions*, Cornell University Press, Ithaca, N.Y., 1962, e.g. pp. 71–74, 77–82, 89, 122–123, and 137–138.

[6] For a brief discussion of what could be done along these lines for philosophical purposes, see my paper, 'A Program and a Set of Concepts for Philosophical Logic', *The Monist* **51** (1967) 69–92. For a survey of the foundational aspects of model theory, see Andrzej Mostowski, *Thirty Years of Foundational Studies* (Acta Philosophica Fennica vol. XVII), Basil Blackwell, Oxford, 1966, ch. 3, ch. 13, and ch. 14. See also J. W. Addison, L. Henkin, and A. Tarski (eds.), *The Theory of Models*, Proceedings of the 1963 International Symposium at Berkeley, North-Holland Publ. Co., Amsterdam 1965 (with a bibliography).

[7] The attempt was made in *Knowledge and Belief* (reference 5 above). The misunderstandings I want to take up here occur in the reviews of this work in *Mind* **75** (1966) 145–149, and *Philosophical Quarterly* **15** (1965) 268–269. Certain other issues which *Knowledge and Belief* has raised are discussed in R. M. Chisholm, 'The Logic of Knowledge', *Journal of Philosophy* **60** (1963) 773–795, in a group of four papers by Chisholm, Castañeda, Sleigh and myself in *Noûs* **1** (1967), no. 1, and in my note 'Knowing Oneself and Other Problems in Epistemic Logic', *Theoria* **32** (1966) 1–13.

[8] Gilbert Ryle, 'Formal and Informal Logic', in *Dilemmas*, Cambridge University Press, Cambridge, 1954.

[9] Cf. Noam Chomsky's methodological remarks in *Syntactic Structures*, Mouton and Co., The Hague, 1957 and in *Aspects of the Theory of Syntax*, The M.I.T. Press, Cambridge, Mass., 1965.
[10] See the two papers of mine referred to in reference 7 above and below pp. 103–4, 161–2.
[11] One way of arguing against the universal applicability of any one particular form of Russell's contextual elimination of definite descriptions is to show that statements in terms of descriptions must sometimes be construed in one of the two ways just indicated and sometimes in the other way. This seems in fact to be the main burden of Keith Donnellan's interesting paper, 'Reference and Definite Descriptions', *Philosophical Review* **75** (1966) 281–304.
[12] In the *Blue Book*, Basil Blackwell, Oxford, 1958, p. 17, Wittgenstein writes, "The study of language-games is the study of primitive forms of language or primitive languages. If we want to study the problems of truth and falsehood, of the agreement and disagreement of propositions with reality, the nature of assertion, assumption, and questions, we shall with great advantage look at primitive forms of language in which these forms of thinking appear without the confusing background of highly complicated thought processes. When we look at such simple forms of language, the mental mist which seems to enshroud our ordinary use of language disappears. We see activities, reactions, which are clear-cut and transparent. On the other hand we recognize in these simple processes forms of language not separated by a break from our more complicated ones. We can see that we can build up the complicated forms from the primitive ones, by gradually adding new forms."

This, indeed, was largely Wittgenstein's program in the *Brown Book*. The fact that he there urges the reader to think not of "the language games we describe as incomplete parts of a language, but as languages complete in themselves, as complete systems of human communication" (p. 81), does not belie that simplicity. This is simply an injunction against bringing in "the confusing background of highly complicated processes of thought" in considering a language-game.

Later, however, Wittgenstein seems to have given up the program expressed in the last few lines of our quotation from the *Blue Book*, at least as an interesting program.
[13] Cf. the highly important discussion of this point and of a number of related ones by H. P. Grice in 'The Causal Theory of Perception', Section III, *Proceedings of the Aristotelian Society, Supplementary Volume* **35** (1961) 121–152; reprinted in Robert J. Swartz (ed.), *Perceiving, Sensing, and Knowing*, Doubleday and Co., New York, 1965, pp. 438–472.
[14] A number of implications due to some of these 'intervening' factors are sometimes grouped together under the heading of 'contextual implications' or 'pragmatic implications'. For a discussion of these, see e.g. Isabel Hungerland, 'Contextual Implication', *Inquiry* **3** (1960) 211–258.

It is important to realize that the more we know (or assume) concerning the purposes which a language (or a certain part of it) is supposed to serve, the more contextual implications we can hope to discover. For a contextual implication from p to q obtains if the act of uttering p (uttering it in a certain way) is pointless, i.e., fails to serve its purpose, unless q is true. The more definite one's purposes are, the more one's words therefore imply contextually. This fact is connected with the usefulness of sharply defined 'language-games' (language-games serving clearly characterized purposes) for the understanding of language.
[15] *Op. cit.* ch. 5, especially pp. 112–123.
[16] *Loc. cit.* especially p. 115.

[17] It is not my purpose in this paper to make any comparisons with the methods of structural linguistics. It may be mentioned, however, that in the present problem I find myself in agreement with the results of J. J. Katz who is prepared to argue that "double negation has as strong support in the semantic structure of English as there is for the simplification in the meaning of *and* in English". See J. J. Katz, 'Analyticity and Contradiction in Natural Language', in *The Structure of Language* (ed. by Jerry A. Fodor and Jerrold J. Katz), Prentice-Hall, Inc., Englewood Cliffs, N.J., 1964, pp. 519–543, especially p. 538.
[18] See the bibliography in Flew, 'Again the Paradigm', referred to in reference 1 above, and the paper mentioned in reference 19 below.
[19] By far the best exposition of what is acceptable in this method is given by Max Black in 'Definition, Presupposition, and Assertion', *Philosophical Review* **61** (1952) 532–550; reprinted in Max Black, *Problems of Analysis*, Cornell University Press, Ithaca, New York, 1954.
[20] A partial list is given in *Knowledge and Belief*, pp. 107–109.
[21] Max Deutscher in *Mind* **75** (1966) 145–149.
[22] It now seems to me that the most straightforward residual meaning, and probably the most common one, is the one in which 'knowing that one knows' amounts to 'being aware that one knows', i.e., sense $(63)_1$ of *Knowledge and Belief*, p. 118. This sense figures in the engaging burlesque of 'ordinary language' philosophizing which Ved Mehta gives in *Fly and the Fly-Bottle* (Penguin edition, 1965, pp. 23–24), to the exclusion of any awareness of the other senses (residual meanings).
[23] *Philosophical Quarterly* **15** (1965) 268–269.
[24] A. R. White, *Attention*, Basil Blackwell, Oxford 1964.
[25] Aristotle was already highly sensitive to the different senses of 'different senses'. See my paper, 'Aristotle and the Ambiguity of Ambiguity', *Inquiry* **2** (1959) 137–151, where I suggest that Aristotle did not take all differences in use to indicate the presence of irreducibly different senses.
[26] *Op. cit.* pp. 17–22, 43–44. Notice that (C.KK*) and (A.PKK*) have the same force (over and above that of the other rules or conditions).
[27] Deutscher, *loc. cit.*
[28] *Op. cit.* pp. 19–20.
[29] *Loc. cit.*
[30] *Knowledge and Belief*, pp. 12–15.
[31] *Op. cit.* pp. 17–21.
[32] Douglas Arner, 'On Knowing', *Philosophical Review* **68** (1959) 84–92.
[33] J. O. Urmson, 'Parenthetical Verbs', *Mind* **61** (1952) 480–496; reprinted in *Essays in Conceptual Analysis* (ed. by A. Flew), Macmillan, London, 1956, pp. 192–212, and in *Philosophy and Ordinary Language* (ed. by C. E. Caton), University of Illinois Press, Urbana 1963, pp. 220–240.
[34] The basic problem here is whether the special functions of parenthetical verbs which are characterized by Urmson make them in some sense non-descriptive so that e.g. the notions of truth and falsehood would not be applicable to the 'parenthetical' remarks made in terms of them. I do not think that the descriptive (indicative) vs. non-descriptive contrast can be drawn very sharply in this area, however, and hence refuse to take the problem at its face value. I have tried to give a glimpse of my reasons in 'A Program and a Set of Concepts for Philosophical Logic', *The Monist* **51** (1967) 69–92, section 2.
[35] Sören Halldén, 'A Pragmatic Approach to Modal Theory', *Proceedings of a Collo-*

quium on Modal and Many-Valued Logics, Helsinki, 23–26 August, 1962, in the series *Acta Philosophica Fennica* **16** (1963) 53–64. Cf. also Sören Halldén, 'A Pragmatic Approach to Modal Logic' in *Filosofiska studier tillägnade Konrad Marc-Wogau 4 april 1962*, (ed. by Ann-Mari Henschen-Dahlquist and Ingemar Hedenius), Uppsala 1962, pp. 82–94.

[36] It still remains to be seen to what extent such requirements of certainty as (C.KK*) can be satisfied in all interesting contexts. Doubts are cast on some such possibilities by the important results of Richard Montague; see his paper 'Syntactical Treatments of Modality', *Proceedings of a Colloquium on Modal and Many-Valued Logics, Helsinki, 23–26 August, 1962*, in the series *Acta Philosophica Fennica*, **16** (1963) 153–167.

We cannot examine here the highly interesting problems which Montague's results pose. It may be pointed out, however, that they do not reflect on the connection which I have argued there is between such conditions as (C.KK*) and the certainty or conclusiveness of 'genuine' knowledge. On the contrary, their importance is apt to be due to this very connection, for in so far as they bring out difficulties in the idea that a notion of knowledge can satisfy (C.KK*) (or some related condition) they demonstrate the difficulty of upholding a standard of 'certain' or 'conclusive' knowledge. It seems to me that epistemologists have not realized the great interest of Montague's results, whatever their adequate interpretation turns out to be.

II. THE LOGIC OF EXISTENCE

EXISTENTIAL PRESUPPOSITIONS AND THEIR ELIMINATION

I. INTRODUCTORY

The notions of existence and necessity have held the interest of philosophers longer than many other problems in the philosophy of logic. The nature of necessity has been debated since the ancient Greeks; and many philosophers have pronounced their opinions on whether 'existence is a predicate'. In this essay, I shall discuss the notion of existence. It serves to prepare the way for the later parts of this book, especially for 'Existential Presuppositions and Uniqueness Presuppositions', where similar methods are applied to the concepts of necessity and possibility.

Modern logic has not so far been quite as helpful in this area as one might expect on the basis of the fact that one of its basic notions is that of an existential quantifier, which is studied in the modern logic of quantification. Unfortunately, logicians have usually been interested primarily or exclusively in the existence of *kinds of* individuals. This is in fact what the usual systems of quantification theory are designed to do. The problem of the existence of individuals *as individuals* has by comparison received only scattered attention. Nevertheless a little more can be said here than earlier writers, including myself, have said so far.

Nor has modern symbolic logic always been as helpful in elucidating the concepts of necessity and possibility as one might hope. This seems to me to be due primarily to the fact that their logic, usually referred to as modal logic, was for a long time studied exclusively by means of syntactical (deductive and axiomatic) methods.[1] Now these methods are not always the best to create philosophical illumination in logic. The methods best suited to increase conceptual clarity are here, as in many other areas of logic, the semantical ones (in the sense of the term in which it has been applied to Carnap's and Tarski's studies). It is not very helpful merely to put one's intuitions into the form of a deductive system, as happens in the syntactical method. They are rarely sharpened in the process. They are usually much sharpened, however, if we inquire into the conditions of truth for the different kinds of sentences that we are dealing with; which is essentially what the semantical method amounts to. In fact, our insights into the notion of truth *simpliciter* and into the closely related notion of satisfiability (truth on some interpretation) are likely to be much richer than our intuitions concerning the problematic concept of *logical* truth. The former are what one is utilizing in the semantical approach, the latter are what one has to resort to directly in the syntactical approach.

II. MODEL SETS

One way (among many) of systematizing our insights into the notion of truth in quantification theory is to deal with what I have called *model sets*.[2] A model set, in short a m.s., is from the intuitive point of view a set of formulas which are all true on one and the same interpretation of the nonlogical constants occurring in them.

In fact, the conditions defining a model set (say μ) are essentially parts of the usual semantical truth-conditions for sentential connectives and quantifiers. They may be formulated as follows:

(C.$\sim$) If p is an atomic formula or an identity, not both $p \in \mu$ and $\sim p \in \mu$.

(C.&) If $(p \& q) \in \mu$, then $p \in \mu$ and $q \in \mu$.

(C.$\vee$) If $(p \vee q) \in \mu$, then $p \in \mu$ or $q \in \mu$ (or both).

(C.E) If $(Ex)\, p \in \mu$, then $p(a/x) \in \mu$ for at least one free individual symbol a.

(C.U) If $(Ux)\, p \in \mu$, then $p(b/x) \in \mu$ for every free individual symbol b occurring in the formulas of μ.

(C.=) If p is an atomic formula or an identity, if $p \in \mu$, if $(a=b) \in \mu$, and if $p(a/b)$ is the same formula as $q(a/b)$, then $q \in \mu$.

(C.self$\neq$) μ contains no formulas of the form $(a \neq a)$.

Instead of (C.self$\neq$) we may alternatively use the following condition:[3]

(C.self=) If b occurs in the formulas of μ, then $(b=b) \in \mu$.

These conditions are self-explanatory except for the fact that it has not been explained what formula is referred to by '$p(a/x)$' in (C.E). This is the formula obtained from p by replacing x everywhere by a. Similar notation is used in the other conditions and frequently in the sequel. In light of this explanation, we can see that the requirement that $p(a/b)= =q(a/b)$ in (C.=) amounts to requiring that q is like p except that a and b have been interchanged at one or more of their occurrences in p.

These conditions suffice if we require (as we may indeed require) that all the formulas we are dealing with have first been reduced to a form in which all the negation-signs have been driven as deep into the formulas as they will go. By means of familiar laws (de Morgan's laws, the law of double negation, the interconnection between the two quantifiers) we can always drive them deeper until their scope is minimal, i.e., until the scope of each consists of a single atomic formula. Notice that even though we shall normally assume that this transformation has been effected, we can go on speaking of such formulas as $\sim(p\,\&\,q)$. When we refer to these formulas, we will simply mean the result obtained by bringing them into our negational standard form.

Some of our conditions have been formulated as conservatively as possible. As far as quantification theory is concerned, they may be strengthened somewhat. For instance, it may be shown that the conditions (C.$\sim$) and (C.=) can in quantification theory be replaced by the stronger conditions,

(C.$\sim$!) If $p \in \mu$, then not $\sim p \in \mu$; and

(C.=!) If $p \in \mu$, $(a=b) \in \mu$, and if $p(a/b)=q(a/b)$, then $q \in \mu$,

respectively, from which the restriction to atomic sentences and identities has been removed. It may be shown that any model set satisfies the additional condition (C.$\sim$!). A proof to this effect is in each case easily conducted by induction on the number of logical constants in p. It may also be shown that a model set μ can always be imbedded in another model set which satisfies (C.=!). This larger set is easily obtained as the closure of μ with respect to the operation of adding a formula which is required to be present by (C.=!).

The main property of model sets is the following: A set of formulas λ is satisfiable (in the usual sense of the word) if and only if there is a model set μ such that $\mu \supseteq \lambda$. If we think in terms of interpreted formulas (sentences), this means that we may think of model sets as descriptions of logically possible states of affairs (possible courses of events, 'possible worlds'). For we undoubtedly want to say that a set of sentences is satisfiable if and only if there is a possible world in which all its members would be true; i.e., if and only if there is a description of a logically possible world which includes all the sentences of λ.

Certain qualifications are needed here, however. First of all, model sets are not (even if we are dealing with interpreted formulas) *complete* descriptions of possible worlds. They are only *partial* descriptions. However, they are large enough to stand on their own feet in the sense of being large enough to show that the state of affairs in question is really possible.

Secondly, it is not quite true to say that imbeddability in a model set is equivalent to satisfiability in the usual sense of the word. It is only equivalent to satisfiability if the empty domain of individuals is admitted on a par with non-empty ones as a domain with respect to which our formulas may be interpreted. In such a domain, every universal sentence is of course true and every existential one false.

In terms of satisfiability, the other central notions may be defined in the usual way. A set of formulas which is not satisfiable is inconsistent. A formula whose negation has an inconsistent unit set is called logically true (valid). A logical truth is a formula which has no conceivable counterexample, we might thus say.

III. EXISTENTIAL PRESUPPOSITIONS

Here we are primarily concerned with the conditions (C.E) and (C.U). If we have a look at (C.U), we can see that this condition is based on important assumptions which it may be useful to avoid.

Ordinarily, $(Ux)p$ is understood to mean 'of each actually existing individual (call it x) it is true that p'. It is undoubtedly possible to understand the universal quantifier in some other way.[4] However, it is not obvious that the alternative readings do not lead into interpretational difficulties. In any case, the meaning of the universal quantifier just explained is clearly its most important sense. And for anyone concerned with the logic of existence it is clearly the meaning he is primarily interested in.

With this observation in mind, we can see that our condition (C.U) is based on the assumption that the free individual symbol b refers to some actually existing individual (or, if we are dealing with an uninterpreted system, behaves as if it did). For clearly from the fact that something is true of all actually existing individuals it does not follow without qualification that the same is true of the individual referred to by b *unless such an individual exists.* And since the condition (C.U) is supposed to be applicable to any b occuring in the formulas of μ, this means that we are in effect assuming that every free individual symbol we are dealing with really refers to an actually existing individual (or behaves as if it did). Everything that can be substituted for a free individual symbol must refer to some individual. Empty singular terms are excluded from discussion. Everything that can be specified by means of a singular term (substitutable for our free individual symbols) must exist.

Since our semantical treatment of quantification is almost equivalent to the traditional deductive systems of quantification theory, all these systems are based on the same presuppositions. Empty singular terms are in the same way ruled out in all of them. We shall call the presuppositions we have thus found *existential presuppositions.*[5]

IV. THE ELIMINATION OF EXISTENTIAL PRESUPPOSITIONS

How can these presuppositions be eliminated? First, should they be eliminated? I am prepared to grant that their elimination does not afford great technical advantages for many of the purposes for which quantification theory is usually employed. Nevertheless, it seems to me that the elimination is desirable in the interests of conceptual clarity. Existential presuppositions in effect prejudge all questions concerning the existence of individuals referred to by singular terms which occur in our model sets or which can be substituted for our free individual symbols. They thus imply the unsatisfactory conclusion that a decision concerning the syntactical status of a term may depend on the decision of the factual question concerning the existence of the individual to which it purportedly refers.

Nevertheless existential presuppositions do not seem to matter greatly as long as we consider only descriptive uses of language in the narrow sense of the term in which descriptive uses of language are contrasted to the use of language, e.g., for the purpose of formulating hypotheses, verifying and falsifying them, making counterfactual statements, etc. The innocence of these presuppositions in descriptive contexts is not very surprising, however, for there is obviously little that can be said by way of pure description of nonexistent individuals.

One might perhaps also hope to limit the substitution-values of free individual symbols to some syntactical category which is restricted narrowly enough to guarantee that existential presuppositions are satisfied by all its members. For instance, it might seem that the category of proper names fills the bill satisfactorily enough. Proper names are often thought of as mere identifying labels attached to individuals we know to exist, without any descriptive content. Hence there does not seem to be any use for them in connection with nonexisting individuals, on which no labels can be pasted.

This view of proper names seems to me oversimplified.[6] In any case, there is a very good case for getting rid of existential presuppositions in contexts in which language is not being used merely descriptively. Here we are primarily interested in modal contexts. When we cease merely to report or to register what is true of the actual world and start to discuss what might not have happened or what could have happened, existential presuppositions soon become awkward. Surely it ought not to be logically inadmissible to try to say something of what might have happened if some particular individual had not existed, e.g., if there had been no Napoleon.

When we consider some other applications of modal logic, the same point emerges even more clearly. One of these is what might be called doxastic logic in which the phrase 'it is believed' or '*a* believes that' takes over the role of the necessity-operator. In such a logic, we certainly want to be able to formulate such sentences as '*a* believes that Ossian really existed' or '*b* believes that he is pursued by the Abominable Snowman' without committing ourselves to the existence of Ossian or of the Abominable Snowman, respectively. In another type of application (tense-logic) the possible states of affairs that are considered are simply states of the world at different moments of time. In order for a singular term not to be empty in any of such state of affairs it must refer to an individual which exists *always*. Surely it would be in vain to look for a syntactically definable category of singular terms such that their bearers *always* exist.

In any case, it seems desirable to investigate the possibility of dropping existential presuppositions. Pending the outcome of such an investigation, the virtues and vices of a logic which tries to dispense with these presuppositions cannot be adequately assessed.

How can we rid our logic of existential presuppositions? First, how do they enter into the semantical system we have formulated? In view of the intuitive meaning of a model set, the gist of these presuppositions may be expressed by saying that the mere presence of a singular term (any substitution-value of a free individual variable) in the description of a state of affairs entails that the individual it purports to refer to really exists in the state of affairs in question. In order to be able to eliminate the presuppositions we want to be able to express the existence of the reference of a singular term (say the term *a*) in such a way that its existence can also be meaningfully denied. In other words, we need a formalization of the perfectly ordinary phrase '*a* exists'.

Can such a formalization be obtained? It may be objected that any such formalization will involve the illicit assumption that 'existence is a predicate'. Fortunately, in a recent note by Salmon and Nakhnikian the standard prima facie objections to treating 'existence as a predicate' have been effectively disposed of.[7] Whether deeper interpretational objections are forthcoming or not, none have been put forward so far; and I doubt very much whether they would at all affect the substance of what we are saying here.

Thus there are no objections to an attempt to find a formal counterpart to the phrase '*a* exists'. Before trying to decide exactly what this formalization might be, let us see what conditions it must satisfy *in any case*. Let us assume that $Q(a)$ is the formal counterpart in question; and let us see how it might be used to eliminate the existential presuppositions on which (C.E) and (C.U) are based.

V. MODIFYING THE CONDITIONS ON QUANTIFIERS

If we drop the assumption that every singular term actually has a bearer, we cannot infer any more from the fact that all actually existing individuals have a certain property that the individual referred to by some given singular term *b* has this property. We can infer this only if we also have the additional premise that the individual in question really exists. Hence (C.U) has to be replaced by a condition in which the existence of the individual referred to by b appears as an additional condition:

(C.U$_q$) If $(Ux)p \in \mu$ and if $Q(b) \in \mu$, then $p(b/x) \in \mu$.

If existential presuppositions are dropped, we likewise have to modify the condition (C.E). It is of course still true that if there are individuals of a certain kind, then at least one individual of that kind must be namable; of that particular individual we can then say that *it* has the property in question. However, if the existential presuppositions are dropped, we must say more of that particular individual: we must add that it really exists. In other words, (C.E) must be replaced by the following stronger condition:

(C.E$_q$) If $(Ex)p \in \mu$, then $p(a/x) \in \mu$ and $Q(a) \in \mu$ for at least one free individual symbol a.

The crucial point is that we have to carry out these modifications *no matter what particular formula will serve as* $Q(b)$. The modified conditions (C.U$_q$) and (C.E$_q$) represent, if I am right, conditions which *any* formalization of the phrase 'b exists' must satisfy. They are the true 'semantical rules' or 'meaning postulates' for the notion of existence.

VI. THE 'PREDICATE OF EXISTENCE' IS DEFINABLE

But if so, we can see what formula will serve as $Q(b)$ *in any case*. It can be shown, on the basis of the modified conditions (C.U$_q$) and (C.E$_q$) plus the unproblematic earlier conditions that

> *the formula* $(Ex)(b=x)$ *or* $(Ex)(x=b)$ *will necessarily have the same logical powers as* 'b exists'.

In order to prove this, it suffices to prove that the two implications

$$Q(b) \supset (Ex)(b = x)$$

and

$$(Ex)(b = x) \supset Q(b)$$

are valid in a quantification theory without existential presuppositions. Their validity of course amounts to the fact that their negations are not satisfiable, i.e., not members of any model set. Hence it suffices to refute the two counterexamples in which these negations are assumed to be satisfiable.

These counterassumptions may be reduced *ad absurdum* as follows:

(A) Assume that $Q(b)\&(Ux)(b\neq x)$ is satisfiable, i.e., that it is a member of some m.s. μ. Then we can argue as follows:

(11)	$(Q(b)\&(Ux)(b\neq x))\in\mu$	
(12)	$Q(b)\in\mu$	from (11) by (C.&)
(13)	$(Ux)(b\neq x)\in\mu$	from (11) by (C.&)
(14)	$(b\neq b)\in\mu$	from (12) and (13) by $(\mathrm{C.U}_q)$

This, however, violates the condition (C.self$\neq$).

(B) Assume that $(Ex)(b=x)\ \&\sim Q(b)$ is satisfiable, i.e. that it is a member of some m.s. μ. Then we can argue as follows:

(21)	$(Ex)(b=x)\ \&\sim Q(b)\in\mu$	
(22)	$(Ex)(b=x)\in\mu$	from (21) by (C.&)
(23)	$\sim Q(b)\in\mu$	from (21) by (C.&)
(24)	$(b=a)\in\mu$	from (22) by $(\mathrm{C.E}_q)$ for some
(25)	$Q(a)\in\mu$	free individual symbol a
(26)	$Q(b)\in\mu$	from (24) and (25) by (C. = !)

Here (23) and (26) violate (C.$\sim$!).

Hence our argument leads to the conclusion that the formal counterpart to the phrase 'b exists' has to be $(Ex)(b=x)$ or some equivalent formula. Nevertheless, all the other equivalent formulas turn out to be more complicated; hence $(Ex)(b=x)$ is the most natural candidate here.

The only step in the arguments (A) and (B) which perhaps calls for further comment is the use of (C.=!) in the step (26) of the argument (B). For certain reasons which we shall not discuss here, it would be better if we could use (C.=) instead of (C.=!). We cannot do so, however, unless we know whether $Q(a)$ is atomic. And in fact it has turned out to be equivalent to a non-atomic formula. Nevertheless, the use of (C.=!) is obviously acceptable. For what its use here amounts to is to say that whenever *a* and *b* are identical and *a* exists, *b* exists too. To this principle there do not seem to be any plausible objections.

VII. LOGIC WITHOUT EXISTENTIAL PRESUPPOSITIONS

The moral of our story so far is clear enough. We can escape the existential presuppositions without any trouble if we change the conditions (C.E) and (C.U) as indicated by (C.E_q) and (C.U_q). However, instead of the redundant primitive predicate $Q(a)$ we can use formulas of the form $(Ex)(x=a)$. The resulting conditions will be called (C.E_0) and (C.U_0), respectively. The condition (C.E_0) may be obtained from (C.E_q) simply by replacing $Q(a)$ by $(Ex)(x=a)$. In the antecedent of (C.U_0) we have to allow any formula of the form $(Ey)(y=b)$ or $(Ey)(b=y)$ to play the role which $Q(b)$ played in (C.U_q).

The result of replacing (C.E) and (C.U) by (C.E_0) and (C.U_0), respectively, is a semantical system of quantification theory different from the original one.[8] The new system will be said to be one *without* existential presuppositions; the old system will be said to be one *with* them. The difference between them affects in the first place the concept of a model set. But since the notion of satisfiability was defined in terms of the notion of a model set, satisfiability in the sense of a system with existential presuppositions has to be distinguished from satisfiability in the sense of a system without them; and the same holds for the other basic notions.

The main difference between the two systems is that in the system with existential presuppositions the mere presence of a free individual symbol *a* in a model set μ is tantamount to the assumption that the individual referred to by *a* exists in the state of affairs described by μ; whereas in a system without the presuppositions this assumption is tantamount to the presence of a formula of the form $(Ey)(y=a)$ or $(Ey)(a=y)$ in μ.

VIII. EMPTY DOMAINS OF INDIVIDUALS EXCLUDED

After having made this point clear, we can also see how an empty domain of individuals can be ruled out as a possible domain of interpretation of the members of a model set in the two systems. In a system with existential presuppositions we have to require that at least one free individual symbol occurs in the formulas of μ. In a system without existential presuppositions we have to require that there is at least one formula of the form $(Ey)(y=b)$ or $(Ey)(b=y)$ present in μ whenever the difference between empty and non-empty domains of individuals is relevant. It turns out that these two requirements can in effect be formulated as follows:

(C.u) If $(Ux)p \in \mu$, then $p(a/x) \in \mu$ for at least one free individual symbol a.

(C.Eself =) $(Ex)(x = x) \in \mu$.

The reason why (C.u) serves the purpose it is cast for (in a system with existential presuppositions) is that the difference between empty and nonempty domains of interpretation is relevant only if there is at least one formula of the form $(Ex)q$ or $(Ux)q$ in μ. The conditions (C.E) and (C.u) together make it sure that in this case there is at least one free individual symbol occurring in the formulas of μ.

In the system without existential presuppositions we shall assume that (C.=) is extended to apply to formulas of the form $(Ey)(y=a)$ and $(Ey)(a=y)$ in addition to atomic formulas and identities. It has already been pointed out that this assumption is clearly justifiable intuitively.

IX. IS EXISTENCE A PREDICATE?

What are the implications of our results so far? First of all, what do they imply concerning the question whether 'existence is a predicate'? Perhaps the main thing we can now see is that the traditional question is equivocal. What I have been arguing is that existence cannot be conceived of as an *irreducible* predicate. Even if we introduce a special predicate $Q(a)$ to express 'a exists', it turns out to be definable in terms of the ordinary existential quantifier. In this sense, existence is expressed by the existential quantifier and by nothing else. Any primitive predicate of existence is *necessarily* redundant if the normal meanings of our other logical concepts are accepted.

If the traditional denial that existence is a predicate is taken to mean that no predicate logically independent of the existential quantifier can express existence, it appears to be correct. But if so, the traditional discussion has been beside the point to some extent. If the burden of the notion of existence is carried by quantifiers in any case, the crucial question will concern the rules governing quantifiers. It is only by studying these rules that we can find whether, and in what sense, existence can perhaps be treated as a predicate or as something like a predicate.

In fact, our examination of the rules that govern quantifiers has not revealed any objections to considering existence as a predicate in a different (weaker) sense of the word. Existence can be a predicate in the sense that it is possible to use a formal expression containing the free individual symbol *a* as a translation of the phrase '*a* exists', without running into any logical difficulties. For the purpose of modifying the conditions (C.E) and (C.U) in the way we found it advisable to modify them it is in fact necessary to have such an expression at our disposal.

For the modification of the conditions (C.E) and (C.U) which turns them into ($C.E_0$) and ($C.U_0$) really gives us something new. It gives us a somewhat richer (more flexible) system in which we can express certain things we could not express before. For instance, we can now meaningfully deny the existence of individuals; formulas of the form $\sim (Ex)\,(x=a)$ are not all disprovable any more. Hence such sentences as 'Homer does not exist' can be translated into our symbolism without any questionable interpretation of the proper name 'Homer' as a hidden description. If anybody should set up a chain of arguments in order to show the non-existence of Homer, we could hope to translate it into our symbolism without too many clumsy circumlocutions. In this sense, the use of an expression for existence is not only possible but serves a purpose. Existence is, if you want, a predicate definable in terms of the existential quantifier.

X. COMPARING THE TWO SYSTEMS

The semantical system obtained by replacing (C.E) and (C.U) by ($C.E_0$) and ($C.U_0$) is weaker than the original system. Those old logical truths that turned on existential presuppositions are not logical truths any more. They can be restored, however, by introducing suitable additional premises which make the underlying existential presuppositions explicit. For instance, $p(a/x) \supset (Ex)p$ used to be a logical truth but is not one any more; however, $(p(a/x) \,\&\, (Ex)(x=a)) \supset (Ex)p$ is still valid (a truth of logic).

This hints at a way of demonstrating the fact we just announced without a proof, viz., the fact that the new system is (in spite of its apparent weakness) richer than the old one. This way is to show that the central logical properties of formulas (logical truth, logical consequence, etc.)

under the old interpretation can be explicitly defined in the new system by means of the logical properties of certain related formulas in the new system. Since all the relevant logical properties are definable in terms of satisfiability, it suffices to show that the satisfiability of a set of formulas in the original sense of the word can be defined as the satisfiability of a certain related set in a system without the existential presuppositions.

In carrying out such a proof, it is important to keep the two systems apart as clearly as possible. I shall refer to a model set defined in the original system *with* existential presuppositions as a model set$^+$, and to a model set in the sense of the new system *without* the presuppositions as a model set$^-$. Similarly, satisfiability in a system with presuppositions will be referred to as satisfiability$^+$, and satisfiability in the sense of the system as satisfiability$^-$; and so on for other notions.

The relation of the two sets of notions may be studied by means of two operations on (arbitrary) sets of formulas. One of them serves to bring out explicitly the existential presuppositions; we shall designate it by *e*. The other serves to throw out everything that does not satisfy the presuppositions; it will be called *f*.

These two operations may be defined as follows: Given λ, $e(\lambda)$ is the set of formulas obtained from λ by adjoining all the formulas $(Ex)\,(x=b)$ where b occurs in at least one formula of λ. These formulas formulate explicitly the existential presuppositions which are implicit in the usual systems. Given λ, $f(\lambda)$ is the set of formulas obtained from λ by omitting every formula which contains at least one free individual symbol a such that no formula of the form $(Ey)\,(y=a)$ or $(Ey)\,(a=y)$ occurs in λ.

The operations *e* and *f* have certain simple properties. The following are among the simplest:

(i) $e(\lambda) \supseteq \lambda$ (ii) $f(\lambda) \subseteq \lambda$

(iii) whenever $\lambda_1 \supseteq \lambda_2$, $e(\lambda_1) \supseteq e(\lambda_2)$;

(iv) whenever $\lambda_1 \supseteq \lambda_2$, $f(\lambda_1) \supseteq f(\lambda_2)$;

(v) $f(e(\lambda)) = e(\lambda)$.

These are all obvious. The following properties are not quite as obvious but perfectly straightforward to verify:

(vi) If μ is a model set$^+$, $e(\mu)$ is a model set$^-$;

(vii) If μ is a model set$^-$, $f(\mu)$ is a model set$^+$,

In fact, if μ satisfies the defining conditions of a model set$^+$, clearly $e(\mu)$ can fail to satisfy the defining conditions of a model set$^-$ for two reasons only: (a) Because (C.E$_0$) or (C.U$_0$) is violated (when applied to some formula already present in μ), for these are the only conditions that are changed when existential presuppositions are given up; or (b) because some new formula introduces violations of some of the conditions. Take the first case first: (C.U) is stronger than (C.U$_0$); hence the presence of any of the old formulas in $e(\mu)$ cannot violate (C.U$_0$). And the only reason why the presence of any formula could violate (C.E$_0$) but not (C.E) is because there are no formulas of the form $(Ey)(y=b)$ or $(Ey)(b=y)$ present in $e(\mu)$ for some b occurring in the formulas of μ. However, the definition of $e(\mu)$ rules this possibility out. This takes care of (a). As to (b), all the new formulas are of the form $(Ex)(x=b)$. Hence the only condition their presence could violate is (C.E$_0$). But by (C.self=) we have $(b=b)\in\mu$; hence (C.E$_0$) is satisfied, too, verifying (vi).

Again, if μ is a model set$^-$, $f(\mu)$ is readily seen to satisfy the conditions (C.$\sim$), (C.&), (C.$\vee$), (C.=) and (C.self=). In order to verify (C.E), assume that $(Ex)p\in f(\mu)$. In view of the definition of $f(\mu)$, this can be possible only if $(Ey)(b=y)\in\mu$ or $(Ey)(y=b)\in\mu$ for each free individual symbol b of p. Because μ satisfies (C.E$_0$), we have $p(a/x)\in\mu$ and $(Ex)(x=a)\in\mu$. But since all the free individual symbols of $p(a/x)$ are the b's and a, we must have $p(a/x)\in f(\mu)$, showing that (C.E) is satisfied by $f(\mu)$. In order to verify (C.U), assume that $(Ux)p\in f(\mu)$ (whence $(Ux)p\in\mu$) and that b occurs in at least one formula of $f(\mu)$. The latter can be the case only if a formula of the form $(Ey)(y=b)$ or $(Ey)(b=y)$ occurs in μ. Since μ satisfies (C.U$_0$), we must have $p(b/x)\in\mu$. Since we had $(Ux)p\in\mu$, for every free individual symbol c of p there must be a formula of the form $(Ez)(z=c)$ or $(Ez)(c=z)$ in μ. Hence the same holds for $p(b/x)$; and by the definition of $f(\mu)$ we therefore have $p(b/x)\in f(\mu)$, verifying (C.U) for $f(\mu)$.

This suffices to prove (vii).

By means of (i)–(vii) we can prove the result we want to prove.

A set of formulas λ is satisfiable$^+$ if and only if $e(\lambda)$ is satisfiable$^-$.

Proof: Assume first that λ is satisfiable$^+$, i.e. that there is a model set$^+$ μ such that $\mu\supseteq\lambda$. Then by (iii) $e(\mu)\supseteq e(\lambda)$. By (vi), $e(\mu)$ is a model set$^-$; hence we see that $e(\lambda)$ can be imbedded in a model set$^-$, i.e. that it is satisfiable$^-$, just as we wanted to show.

In order to prove the other half of the equivalence, assume that $e(\lambda)$

is satisfiable$^-$, i.e. that there is a model set$^-$ μ such that $\mu \supseteq e(\lambda)$. Then by (iv) $f(\mu) \supseteq f(e(\lambda))$. By (v) and (i) $f(e(\lambda)) = e(\lambda) \supseteq \lambda$, and hence $f(\mu) \supseteq \lambda$. But by (vii) $f(\mu)$ is a model set$^+$, hence λ is satisfiable$^+$, as we wanted to prove.

Theorem I shows that satisfiability$^+$ can be defined in a simple way in terms of satisfiability$^-$. In an important sense, the old system can therefore be interpreted as a subsystem of the new one. Since no simple result in the other direction is forthcoming, the new system is in fact richer than the old one.

XI. OTHER CANDIDATES FOR THE ROLE OF A 'PREDICATE OF EXISTENCE'

My approach to the problems of individual existence may be illustrated by comparing it with certain other approaches. The main problem here is the choice of the formula to serve in the role of $Q(b)$ as the 'predicate of existence'. The favourite earlier candidate for this role seems to have been the formula $(b=b)$.[9] Using it for this purpose necessitates the rejection of the condition (C.self$\neq$) (and of the condition (C.self=)), for the main point in using the predicate of existence is to be able to deny existence to individuals without contradiction. But if we reject it, the formula $Q(b) \supset (b=b)$ will not be valid. Nor is the converse implication $(b=b) \supset Q(b)$ validated by our other conditions, including the modified conditions (C.E_0) and (C.U_0). In other words, there is no positive evidence for the equivalence $(b=b) \equiv Q(b)$ which equates $(b=b)$ with the predicate of existence $Q(b)$, as there was for the identification of $Q(b)$ with $(Ex)(x=b)$. In so far as our modified conditions formulate the properties of the concepts with which they deal exhaustively, the use of $(b=b)$ as a predicate of existence is therefore entirely groundless.

Moreover, if we nevertheless push the formula $(b=b)$ into the role of a predicate of existence, difficulties will ensue. The necessity of having to give up (C.self$\neq$) is already awkward to motivate. The interpretation of the equivalence $Q(b) \equiv (b=b)$ which formally identifies $(b=b)$ with the predicate of existence $Q(b)$ is also rather difficult. Do those who favor this approach want to say that 'Homer is Homer' implies that Homer existed? Or that we have to deny that Hamlet was identical with Hamlet in order to be able to deny that he really existed? I cannot associate any clear sense with these statements, and I cannot see any reasons for incorporating them into one's logical system.

What is worse, the condition (C.=) cannot stand up any more either. For surely we may want to assert and to deny identities between individuals without being committed to their existence. We might e.g. want to disprove Homer's existence by considering several possible identifications of Homer with other individuals. But if we assert 'Homer$=a$', for any free individual symbol a whatever, no matter whether it is assumed to refer to anything or not, then we have by (C.=) 'Homer=Homer', i.e., we are committed to Homer's existence. Thus (C.=) presumably will have to be changed somehow. Whatever the changes are, they are likely to be unnatural. For instance, we cannot any more uphold both symmetry and transitivity, for they would together give us again 'Homer= Homer' from 'Homer$=a$'. The only way of avoiding such radical revisions would be to deny that a non-existent individual can ever be truly said to be identical with any individual, existent or non-existent.

In general, it seems unnatural to try to find room for changes in one's ways of dealing with the notion of existence by changing the rules which govern the notion of identity. What have these two to do with each other in the first place? Why cannot we change the interpretation of the one without having to change the conditions governing the other? In contrast, it is only natural that a change in our ways of dealing with existence will necessitate changes in the conditions governing the logical behavior of quantifiers, for these (especially the existential quantifiers) of course are concerned with the notion of existence.

The popularity of $(b=b)$ as a candidate for the role of our 'predicate of existence' $Q(b)$ probably derives from a misguided application of Russell's theory of definite descriptions. It is thought that a proper name or other free singular term behaves, at least in the contexts where it cannot be assumed to have a reference, like a definite description $(\imath x)\ B(x)$ derived from some predicate expression $B(x)$. Indeed, if this definite description is allowed to replace b, the identity $(b=b)$ becomes (on Russell's theory) equivalent to

(31) $$(Ex)\,(B(x)\,\&\,(Uy)\,(B(y) \supset x=y))$$

which is also equivalent (on the same theory) to what becomes of

(32) $$(Ex)\,(x=b)$$

when the same replacement is made. Hence $(b=b)$ seems to play the role of a predicate of existence quite as well as (32). This impression may even be heightened by observing that the second member of the conjunction occurring in (31) may be taken to be true in virtue of the meaning of B. (Only predicate expressions which are satisfied by at most one individual give rise to definite descriptions which can serve to replace a proper name.) Hence (31) is essentially equivalent to the formula $(Ex)B(x)$, i.e. essentially a statement that $B(x)$ is not empty.

These reasons in favour of $(b=b)$ as a predicate of existence are completely illusory, however. They are due to the fact that certain existential presuppositions have already been built into Russell's theory of definite descriptions (presuppositions of a somewhat different kind from the ones discussed so far). These presuppositions are shown by the contextual definitions which are basic in Russell's theory; a typical example of them is constituted by the contextual definitions which the following schema enables us to make:

$$\text{(33)} \qquad \phi((\imath x)B(x)) \equiv (Ex)\,(\phi(x)\,\&\,B(x)\,\&\,(Uy)\,(B(y) \supset x=y)).$$

This schema shows that for a Russellian all use of definite descriptions contains implicit statements of existence; these are the existential presuppositions incorporated in Russell's theory of definite descriptions that I mentioned.

As I have pointed out elsewhere,[10] we can get rid of these presuppositions by using instead of Russell's contextual definitions such contextual definitions which are illustrated by the following equivalence:

$$\text{(34)} \qquad (a=(\imath x)B(x)) \equiv (B(a)\,\&\,(Ux)\,(B(x) \supset x=a)).$$

If (34) instead of (33) is the basis of our theory of definite descriptions, a substitution of $(\imath x)B(x)$ for b in (32) gives rise to a formula which is still equivalent to (31) and which implies $(Ex)B(x)$. Nevertheless the same substitution in $(b=b)$ gives rise to a formula which is *not* any more equivalent to (31) and which does not any more imply $(Ex)B(x)$. In general, $(b=b)$ does not imply any existential statements any more even if b is replaced by a definite description. Hence the apparent success of $(b=b)$ in the role of a predicate of existence is really due to the presence of existential presuppositions, either in the original form (which we are here trying to eliminate) or in the form of assumptions tacitly built into Russell's theory of definite descriptions.

XII. QUINE'S THESIS

Our results may also throw some light on Quine's famous thesis that 'to be is to be a value of a bound variable'.[11] Commentators have been puzzled by this *dictum*, and not without good reasons. One of these reasons is that the reference to *bound* variables in Quine's thesis seems to be unwarranted. Suppose something or someone, say the individual referred to by the singular term t, is a value of a *free* variable. Now the principle which is known as existential generalization is valid in ordinary systems of quantification theory, giving us as a special case the validity of

$$(a = a) \supset (Ex)(x = a),$$

where a is a free individual variable. Since the individual referred to by t is a value of a free variable, t must be substituted for free individual variables. From the formula just displayed we thus obtain by substitution

$$(t = t) \supset (Ex)(x = t).$$

Since $(t=t)$ is clearly true, we obtain $(Ex)(x=t)$ by *modus ponens*. But what this sentence says is that the individual referred to by t is identical with one of the values of the bound variable x, i.e., is a value of a bound variable.

In short, in ordinary systems of quantification theory, every value of a *free* variable is also a value of a *bound* variable. Hence restricting Quine's *dictum* to bound variables seems unnecessary. In fact, Quine himself occasionally drops this restriction and formulates the principle in terms of variables in general.[12]

But if the *dictum* is formulated in terms of variables in general, then Quine's principle is not his any more. In the form in which no distinction is made between free and bound variables, the principle was already put forward by K. Ajdukiewicz in his doctoral dissertation *Z metodologii nauk dedukcyjnych* (Lwów 1921).

Leaving questions of history aside, our observations in this paper suggest a way of arguing that Quine's *dictum* is really justifiable and important, and that even the controversial restriction of its application to bound variables is defensible and interesting. No matter how Quine himself originally conceived of the meaning of the *dictum*, it seems to me that by far the most important way of interpreting it is to take it to say that formulas of the form $(Ex)(x=a)$ serve as a formalization of the common-sense phrase 'a exists'. For what $(Ex)(x=a)$ says is that the individual referred to by a is identical with one of the values of the bound variable x; and being identical with one of its values is obviously the same as simply *being* one of the values. In a couple of earlier studies, I have suggested that Quine's thesis is correct in the weak sense that formulas of this form *can* serve this purpose.[13] In the present paper, I have argued for a stronger thesis. I have argued that these formulas not only *may* serve this purpose but *must* do so in the sense that they are (up to a logical equivalence) the *only* formulas which can serve this purpose. Even if the existential presuppositions on which usual systems of quantification theory are based are given up, sentences of the form $(Ex)(a=x)$ will necessarily have the logical force of the sentence 'a exists'. I have thus proved that Quine's thesis is correct in a rather strong sense.

At the same time, the elimination of the existential presuppositions shows that the word 'bound' in Quine's *dictum* is indispensable. The argument by means of which I sought to suggest that it perhaps is dispensable was based on the principle of existential generalization. Now this principle is clearly the first and foremost principle that goes by the board as soon as the existential presuppositions are relinquished. Hence the argument for redundancy applies only to traditional formulations of quantification theory. In fact, it is readily seen that in a system without existential presuppositions existence is no longer tantamount to being a value of a free variable. What happened when the presuppositions were given up was just that empty singular terms were admitted as substitution-values of *free* individual variables, although of course not of bound individual variables. For the first time, the word 'bound' in Quine's *dictum* is therefore not redundant any more.

Nevertheless it seems to me that some of the best known formulations of Quine's thesis are somewhat misleading. There is nothing special about bound variables which 'commits' us to the existence of certain entities while the use of other symbols does not. What commits us to the existence of individuals is of course the existential assertions that we make explicitly or implicitly.[14] What is true about Quine's thesis is in my view that each of these ways of making 'existential commitments' (existential assertions) is logically equivalent to asserting the existence of a suitable value of a bound variable. It is not that we make existential commitments only when we use bound variables; rather, the fact is that whenever we make them we might as well use existential quantifiers and bound variables. To this result we are led (if my arguments have been correct) by certain fairly obvious features of our logic of existence and universality; we are committed to it, it might perhaps be said, by our own ways with these notions.

XIII. SOME MORALS OF OUR STORY

It may be useful to formulate explicitly certain general precepts for the kind of explication of the meaning of logical constants which we have been carrying out. They are partly directions which have guided us in our analyses and partly *desiderata* which we have been able to achieve.

(i) The meaning of a logical constant is best brought out by the semantical rules which govern it.

Comment: The conditions defining a model set are essentially such rules.

(ii) Insofar as the meaning of a logical constant is independent of the meanings of others, the rules governing it should be formulated independently of the rules governing the others.

Comment: In the conditions defining a model set, each logical constant we are considering occurs in one condition only, with the only exception of the identity sign.

This requirement makes it also possible to change the rules governing one constant while leaving the rules governing the others intact. This is just what we are able to do in changing the rules for quantifiers so as to rid ourselves of the existential presuppositions.

(iii) When a change in the rules for some logical constants is made desirable by certain presuppositions or other hidden assumptions, then the change in the rules should be such that the assumptions will thenceforward be represented by explicit premises.

Comment: This is just what we did when we introduced the expression $Q(b)$. It was calculated to serve as the explicit premise which brings to the open the existential presuppositions.

Afterwards, we found that we could also satisfy the following requirement:

(iv) When a logical constant is reinterpreted, no new logical constants should be introduced for the purpose.[15]

This requirement was satisfiable in that a familiar old expression turned out to be able to play the role of $Q(b)$.

It will turn out that the same precepts will guide us to a solution of certain central problems in modal logic.

REFERENCES

[1] In this respect, a profound change has been brought about by the work of Stig Kanger and Saul Kripke. See Kanger, *Provability in Logic, Stockholm Studies in Philosophy*, vol. I, Stockholm 1957; Kanger's papers in *Theoria* **23** (1957) 1–11, 133–134, and 152–155; Saul Kripke, 'A Completeness Theorem in Modal Logic', *Journal of Symbolic Logic* **24** (1959) 1–14; Saul Kripke, 'Semantical Considerations on Modal Logic', *Acta Philosophica Fennica* **16** (1963) 83–94; Saul Kripke, 'Semantical Analysis of Modal Logic I', *Zeitschrift für mathematische Logik und Grundlagen der Mathematik* **9** (1963) 67–96; Saul Kripke, 'Semantical Analysis of Modal Logic II', in *The Theory of Models*, Proceedings of the 1963 International Symposium in Berkeley (eds. J. W. Addison, L. Henkin and A. Tarski), Amsterdam 1966, pp. 206–220. Cf. also my papers, 'Modality and Quantification', *Theoria* **27** (1961) 119–128, and 'The Modes of Modality', *Acta Philosophica Fennica* **16** (1963) 65–81 (present volume, pp. 57–70 and 71–86, respectively).

[2] The technique of model sets was explained in my work, 'Form and Content in Quantification Theory', *Acta Philosophica Fennica* **8** (1955) 7–55. They have been used in the papers of mine mentioned in ref. 1, in my book, *Knowledge and Belief*, Cornell University Press, Ithaca, N.Y., 1962, and in my paper, 'Quantifiers in Deontic Logic', *Societas Scientiarum Fennica, Commentationes hum. litt.* **23** (1957), no. 4.

[3] The two conditions yield somewhat different classes of model sets. The difference is inessential, however, for we shall see that the crucial thing is imbeddability in a model set. Now each model set satisfying the condition (C.self≠) can easily be imbedded in a model set satisfying (C.self=), and *vice versa*; hence the difference does not matter.

[4] Cf. Ruth Barcan Marcus, 'Interpreting Quantification', *Inquiry* **5** (1962) 252–259. Although the interpretation Mrs. Marcus offers of quantifiers is a highly interesting one in its own right, it seems to me that it is not relevant to our problem concerning the logic of the notions of (actual) existence and (actual) universality. Although we are willing to admit empty singular terms as substitution-instances of our *free* variables, they have to be excluded from the range of possible substitution-instances for *bound* variables. In a sense, our main problem is just to find suitable ways of doing so.

[5] I have studied them briefly in my paper, 'Existential Presuppositions and Existential Commitments', *Journal of Philosophy* **56** (1959) 125–137.

[6] A more realistic account of proper names has been given by John R. Searle in his paper, 'Proper Names', *Mind* **67** (1958) 166–173. Searle argues that the 'descriptive presuppositions' on which the use of proper names like 'Tully' and 'Cicero' is based may be such that an identity between names is synthetic. By the same token, some of these presuppositions may conceivably fail to be fulfilled, whence an existential statement involving a proper name may be synthetic.

[7] G. Nakhnikian and W. Salmon, '"Exists" as a Predicate', *Philosophical Review* **66** (1957) 535–542.

[8] The same system was formulated in a different way in my paper, 'Existential Presuppositions and Existential Commitments' (cf. ref. 5). An equivalent system was independently put forward by H. Leblanc and T. Hailperin in 'Non-designating Singular Terms', *Philosophical Review* **68** (1959) 239–243.

[9] It is used as the predicate of existence by Nakhnikian and Salmon, and it is also mentioned as one possible candidate for this role by Timothy Smiley in his interesting paper, 'Sense without Denotation', *Analysis* **20** (1959–60) 125–135.

[10] See 'Towards a Theory of Definite Descriptions', *Analysis* **19** (1958–59) 79–85. This paper needs a further specification, however, restricting the use of (34) to those cases in which *a* is not a description.

[11] W. V. Quine, *From a Logical Point of View, 9 Logico-Philosophical Essays*, Harvard University Press, Cambridge, Mass., 1961, 2nd ed., revised. (See especially essays 1 and 6.) W. V. Quine, *Word and Object*, M.I.T. Press, Cambridge, Mass., 1960.

[12] See e.g., *From a Logical Point of View* (ref. 11), p. 13.

[13] In 'Existential Presuppositions and Existential Commitments' (ref. 5) and subsequently in other works.

[14] Cf. Noam Chomsky and Israel Scheffler, 'What Is Said To Be', *Proceedings of the Aristotelian Society* **59** (1958–59) 71–82.

[15] In contrast, Smiley's interesting suggestions (ref. 9) turn on the use of an additional primitive.

ON THE LOGIC OF THE ONTOLOGICAL ARGUMENT

Some Elementary Remarks

It is much harder than one might first suspect to see what is wrong – if anything – with the ontological argument, in some of its variants at least. By way of criticism, it is often said that the argument fails because 'existence is not a predicate'. However, there are senses – and what is more, senses other than the purely grammatical one – in which existence clearly *is* a predicate. It is sometimes said that existence is not the kind of property that can be included in the essence of anything; but the reasons for saying so are far from clear, and the notion of essence is a notorious mess in the best of circumstances. One might suspect that something goes wrong with the logic of definite descriptions in the modal contexts involved in the argument; but I shall try to reconstruct some of the most important aspects of the ontological argument in terms having little to do with ordinary modalities and nothing whatsoever with definite descriptions. In fact, the independence of the essential features of the ontological argument from the theory of definite descriptions ought to be clear enough without much detailed argument. If what we are trying to do is to establish that there exists a unique being "than which nothing greater can be conceived" – in short, a unique supremely perfect Being – surely the great difficulty is to show that there exists *at least one* such being, whereas we can face the problem of uniqueness with relative calm.

Furthermore, it has been complained that the notion "being greater than anything else that can be conceived of" and the notion of supreme perfection are unclear. More than that, it is sometimes suggested that they are systematically ambiguous – that they make no sense until it has been specified in what respect greatness or perfection is to be measured. Certainly, greater evil or more perfect vice cannot be what is meant – but even if there be no such things as these, what precisely *is* meant?

Yet a straightforward answer to this question is forthcoming. What is at stake is surely greatness or perfection with respect to existence. It does not take a neo-Platonist to agree that the greatest or most supreme being intended in the argument is certainly one whose powers of existing are maximal or whose mode of being is, as existence qua existence goes, supremely perfect.

Can we express in some reasonable way that some x is such a being – at least one such being? There are very natural-looking candidates for this task. One thing we can do is to say that x is an existentially perfect being, in short $Pr(x)$, if and only if it exists, provided that anything at all exists:

$$(1) \qquad Pr(x) \equiv (Ez)(z = z) \supset (Ez)(z = x).$$

Here I have rendered 'x exists' by '$(Ez)(z = x)$.' The reasons why one has to do so are given in another paper of mine.[1]

It is to be noticed that the role of x in (1) is that of a placeholder for singular terms, not one of a bindable variable.[2]

Prima facie at least, $Pr(x)$ as defined by (1) seems to express accurately and fully the idea that x is existentially the most perfect being (or one such being): nothing at all can exist without this x also existing – or, if the expression is allowed, all the other beings are existentially dependent on x.

It is easily seen that, provided the world is not completely empty, such a perfect being must exist:

$$(2) \qquad (Ez)(z = z) \supset (Ex)\, Pr(x)$$

is *logically true*. Hence a version of the ontological argument seems to possess perfectly good logical validity after all. Moreover, it can be shown that the logical truth of (2) does not depend on any hidden existential presuppositions.[3]

I strongly suspect that the logical truth of (2) (together with a number of related truths) is an important part of the tacit and half-understood reasons why the ontological argument is so perennially tempting.

Though logically true, this 'ontological argument' is useless for the purposes which it was calculated to serve, as one can see simply by rewriting (2) or

$$(Ez)(z = z) \supset (Ex)((Ez)(z = z) \supset (Ez)(z = x))$$

by means of an elementary transformation into

$$(3) \qquad (Ex)(x = x) \supset ((Ez)(z = z) \supset (Ex)(Ez)(z = x)).$$

It is patent that (3) is completely vacuous. In fact, (3) shows at once that *any* existing individual will serve as the desired kind of x whose existence is asserted in the consequent of (2). Notice that if a value of z exists which makes $(Ez)(z=z)$ true, then *this same* individual serves as the value of both x and z which makes $(Ex)(Ez)(z=x)$ true. Hence the existence of no particular entity is established by the logical truth of (2).[4]

This may be thrown into sharper focus by observing that (2) is to all practical purposes an instance of the schema

(4) $$(Ez)(z=z) \supset (Ex)((Ez)A(z) \supset A(x)).$$

The logical truth of (4) may seem impressive to an uninitiated; it e.g. seems to imply that, given any problem whatsoever, there is a man who is able to solve it if anybody is (provided the universe is not empty). Yet the trick involved is exposed in many elementary logic texts. If at least one man can solve a problem, *any* such man serves as an instance of the kind of x claimed to exist in the consequent of (4).

The failure of (2) to give us a characterization of the kind of existentially perfect being we are looking for is not accidental: no other attempted characterization would have fared any better. Any condition on x that you may care to formulate in the sole terms of bindable variables, quantifiers, connectives, identity, and a predicate of existence, will be logically equivalent either to a vacuous predicate which applies to all existing individuals, to a contradictory one, or to a simple numerical condition on one's domain of individuals (e.g. to the condition that x is the only individual, or that there are at least two other individuals, or to some such thing). This can be proved in the treatment of presupposition-free logic which I outlined in my 1966 *Monist* paper (mentioned in the first reference of the present paper). I argued there that $(Ez)(z=b)$ will always do the duty for the expression 'b exists'; hence the reference to a special predicate of existence can be omitted. The rest can be proved formally by a simple argument.

Likewise, one can argue that any characterization of a kind of individual x in terms of given predicates, bindable variables, quantifiers, connectives, identity, and a predicate of existence can always be replaced by a straightforward description without any special predicate of existence and with the identity-relation occurring only in the following contexts: $(Ey)(y \neq z_1 \& y \neq z_2 \& \ldots \& y \neq z_k \& F(y, z_1, \ldots, z_k))$; and $(y)(y=z_1 \vee y=z_2$

$\vee \ldots \vee y = z_k \vee F(y, z_1, z_2, \ldots, z_k))$. In other words, if an exclusive interpretation of quantifiers is used, no identity signs are needed.[5]

This simple result shows clearly what is true in the misformulated cliche that "existence is not a predicate". Existence *is* a predicate. There is no grammatical or logical harm whatsoever in treating it like one, and in using it for the purpose of characterizing different kinds of individuals. What is peculiar about it is that it is redundant for all descriptive purposes. If this is what is meant by statements of the well-known Kantian kind, to the effect that "by whatever and by however many predicates we may think a thing – even if we completely determine it – we do not make the least addition to the thing when we further declare that this thing is", then such statements are completely correct. Let us notice, moreover, that this correctness does not in any way depend on the problematic concept of essence.

Does the ontological argument fare any better if we introduce modal operators? Let us examine the situation. In order to be as clear as possible of the different assumptions involved, let me couch my discussion in terms of the epistemic operator 'it is known that', in short, K. The dual operator P will have the force of saying 'for all that is known, it is possible that'. I prefer working with these because their intuitive meaning, and as a consequence most of their semantical behavior, is much clearer than e.g. those of logical modalities 'it is logically necessary that' and 'it is logically possible that'.[6] However, I expect that essentially the same points as I shall proceed to make in terms of the epistemic modalities K and P can be made in terms of logical modalities, in so far as they are viable at all.

In terms of K and P it seems to be easy to formulate a characterization of an existentially perfect being, say x. Of any such being (if any) it is surely known that if anything exists, it will do so too:

(5) $$K((Ez)(z = z) \supset (Ez)(z = x)).$$

Let us abbreviate this with $Pr'(x)$. Can we then prove that such beings exist, i.e. can we prove the following:

(6) $$(Ex)\,Pr'(x)?$$

Proving (6) would mean showing that its negation cannot be a member of any 'model set', i.e. of any consistent description μ of a possible world.[7]

Let us assume that it is, and see whether anything impossible results:

$$(7) \qquad (x)\, P\,((Ez)(z=z)\,\&\,(z)(z \neq x)) \in \mu.$$

Nothing follows from (7) unless we have some term b at hand of which it is known whom it refers to:[8]

$$(8) \qquad (Ez)K(z=b) \in \mu.$$

By well-known principles, (7) and (8) imply

$$(9) \qquad P((Ez)\,(z=z)\,\&\,(z)\,(z \neq b)) \in \mu,$$

hence[9]

$$(10) \qquad (Ez)(z=z) \in \mu^*$$

and

$$(11) \qquad (z)(z \neq b) \in \mu^*$$

for some alternative state of affairs μ^*. Furthermore, we must have[10]

$$a = a \in \mu^*$$
$$(Ez)\,(z=a) \in \mu^*$$

for some a. Nothing further follows, however, unless we assume that (8) implies

$$(12) \qquad (Ez)(z=b) \in \mu^*.$$

However, if this is assumed, (12) will contradict (11).

If the two assumptions that were mentioned in the course of the argument are made, our new 'ontological proof' thus succeeds. What does this result show?

The validity of the second assumption which enabled us to carry out the proof is easily seen to be tantamount to the validity of the implication

$$(13) \qquad (Ex)K(x=a) \supset K(Ex)(x=a).$$

In plain English, the assumption says that we can know *who* someone is only if we know *that* he exists.[11] This assumption has a great deal of initial plausibility; so much so that I assumed it in *Knowledge and Belief* as a valid principle.[12] However, for several concurrent reasons I have come to consider it as invalid. Now we can see that the question of its validity is of a considerable interest to the evaluation of the ontological argument. If the validity of (13) is assumed, a version of the ontological argument can be carried out.

What are my reasons for rejecting the validity of (13)? An illustration of them is obtained by pointing out that if it is assumed, we could not formulate in all the natural ways we might want to use such perfectly natural statements as 'there is someone who is not known by a to exist'. One possible formalization of this statement is

$$(14) \qquad (Ex) \sim K_a(Ey)\,(x = y)$$

the negation of which is easily seen to be implied by (13) (or which can be shown to be contradictory by means of the assumption that enabled us to vindicate a version of the ontological argument). Analogous remarks pertain to such statements as

$$(Ex)\, B_a \sim (Ey)\,(y = x)$$

and

$$(Ex) \sim B_a(Ey)\,(y = x).$$

(This does not quite settle the matter, however, as I shall try to explain in a supplementary note appended to the present paper.)

Hence there seem to be good reasons for denying the validity of (13), and hence our version of the ontological argument goes by the board.

It is important to realize, however, that my rejection of the validity of (13) is not due to a desire to make the ontological argument invalid, although the rejection results in the overthrow of a version of the argument. By considering the argument given above, it can be seen that it does not establish the desired result in any case, independently of whether or not the validity of (13) is assumed.

This is seen by recalling that an additional premise of the form $(Ex)\,K(x=b)$ (i.e. 'it is known who b is') was needed anyway, for some free singular term b. This term was needed as a substitution-value of x in (7). In other words, this term was brought in in order to provide a counter-argument to the assumption that a perfect being (in the sense of a being satisfying $Pr'(x)$) does not exist. (In fact, the consideration of all the other singular terms is seen to be beside the point in our argument.) To all intents and purposes, we thus had to assume that it is known who the perfect being is before we could prove that He exists. But we could prove His existence only by assuming the validity of the principle that knowing who *b* is presupposes knowing that *b* exists. Hence, if this principle is assumed, it is strictly circular to assume the additional premise (8). Hence our reconstrued 'epistemic' version of the ontological argument must be said to fail already on account of this circularity.

Moreover, if the validity of (13) were assumed, the characterization of God as the most perfect being (in the sense of a being satisfying $Pr'(x)$) would be quite unnecessary, for it is seen from the above argument that no use is made in it of the antecedent of the implication for which $Pr'(x)$ is a shorthand. (This point is similar to the point made earlier that the validity of (2) does not go to show the existence of any particular being.) The whole force of the argument would reduce to saying that since it is known who God is, He is known to exist. And merely saying this without further explanation is scarcely taken by anyone to amount to an argument for God's existence, although it is perhaps not too far from the traditional idea that our having an adequate idea of God is sufficient to prove His existence.

Our argument was couched in terms of one particular attempt to define God as the existentially most perfect being – "a being than which a greater (existentially greater!) cannot be conceived". It can be shown, however, that no other characterization along similar lines can succeed any better. By reviewing all the different characterizations that one may try to give of an existentially perfect being – or of any being, for that matter – in the sole terms of the predicates of identity and existence, the concept of knowledge, quantifiers, and propositional connectives, one can see that no one of them makes an essential difference to our attempts to prove the existence of a being so characterized. I shall not try to prove this result here, nor state it more explicitly. Suffice it to say that it is a straightforward consequence of the adequacy of any reasonable system of epistemic logic that I know of. It extends to epistemic logic the result which (I suggested earlier) is the gist in the idea that existence is not the kind of attribute which can constitute the essence of any one thing.

How close does our attempted reconstrual of the ontological argument come to the real thing? The argument was formulated by Anselm in terms of "existence in the mind" vs. "existence in reality". This distinction is often explicated in terms of possible existence vs. actual existence. In this note, I have in effect replaced this explication by another one, to wit, by a distinction between something's (say *b*'s) existing in the mind (say in the mind of *a*) in the sense of *a*'s knowing who *b* is, and *b*'s existing actually.

This way of going about seems to me preferable for several reasons. First of all, it appears to come much closer to Anselm's language of *understanding* or *being able to conceive of* "a being than which a greater cannot be conceived" than any talk of what is possible. Second, the notions of (conceptual) possibility and (conceptual) necessity are notoriously obscure; their characteristics have been debated back and forth. In comparison, the idea of knowledge, including the idea of *knowing who* someone is, is a commonplace, however difficult its full analysis is likely to be. Hence we are apt to be much more knowledgeable about this concept than about the somewhat artificial philosophers' notions of conceptual necessity and conceptual possibility. In particular, we have a much better grasp of the idea of *knowing who* someone is than we have of the philosophical concepts of essence and conceptual possibility. It may in fact be said that the concept of essence was in our sample argument replaced by the concept of knowing who someone is.

Perhaps this does not make much difference, however, for it seems to me that in so far as one can build a satisfactory theory of (conceptual) necessity, it will be in the relevant respects sufficiently similar to the logic of knowledge to enable us to say essentially the same things about our chances of reconstructing the ontological argument in terms of ordinary modal logic as we already said about these chances in epistemic logic (the logic of knowledge). Gaunilo, Aquinas, and Kant thus appear to have been shrewder – or perhaps merely sounder – logicians than St. Anselm and Descartes.

SUPPLEMENTARY NOTE

As was mentioned in the text, I have come to give up the conditions (C.EK=) and (C.EK=)* of *Knowledge and Belief.* I did this initially as a response to certain critical remarks. Although I still think that the two conditions have to be given up, I have meanwhile realized that most of these criticisms fall much short of establishing this. For instance, Hector-Neri Castañeda has claimed that the following consistent statement of ordinary language cannot be consistently expressed in my symbolism:[13]

(15) there is a person such that Jones does not know that the person in question exists.

The formulation

(16) $(Ex) \sim K_{\text{Jones}}(Ey)(x = y)$,

which appears to be the most straightforward rendering of (15) in my symbolism, is indeed inconsistent unless (C.EK=) and (C.EK=)* are given up.

However, this is not the whole story. I have indicated repeatedly how a distinction can be made between what is said of the reference of a singular term, 'whoever he is or may be', and between what is said of the definite individual to which a singular term in fact happens to refer.[14] If the person whose existence is asserted in (15) is Smith, then the natural formulation of

(17) Jones does not know that Smith exists

will surely construe it as a statement about the individual in question (i.e. about Smith, the flesh-and-blood person). In other words, (17) is really of the form

(18) $(Ex)(x = \text{Smith}) \& \sim K_{\text{Jones}}(Ey)(y = x)$

or

(18)* $(x)(x = \text{Smith} \supset \sim K_{\text{Jones}}(Ey)(y = x)$.

But if this is so, the natural translation of (15) surely is not (16) but rather

(19) $(Ez)(Ex)(x = z \& \sim K_{\text{Jones}}(Ey)(y = x))$

or

(19)* $(Ez)(x)(x = z \supset \sim K_{\text{Jones}}(Ey)(y = x))$.

Of these (19)* is not inconsistent even if the two critical conditions are presupposed. Although (19) is inconsistent if (C.EK=) or (C.EK=)* is assumed, its (conditional) inconsistency is not altogether surprising. It merely reflects the plausible (but misleading) basis of Castañeda's criticism: If Smith's existence is not known to Jones, Smith cannot be identical with one of the individuals whose identity is known to Jones. But this does not make (19)* any worse a translation of (15), and Castañeda's objection is therefore rebutted even if the critical conditions are assumed to be satisfied.

REFERENCES

[1] Jaakko Hintikka, 'On the Logic of Existence and Necessity. I Existence', *The Monist* **50** (1966) 55–76; reprinted, with the title 'Existential Presuppositions and Their Elimination', in the present volume, pp. 23–44.

[2] This remark is necessitated by the difference in logical behavior between free singular terms and bound (or bindable) variables which they evince as soon as we give up the 'existential presuppositions' to the effect that each free singular term refers to some individual. For details, see the paper referred to above.

[3] This is easily seen by means of the technique employed in the paper cited above.

[4] The basic difficulty about the ontological argument is thus not so much that it is invalid, but that it only appears to establish what it seems to prove, and that the argument is not readily seen to be the tautology it is. Its critics have for this reason levelled their objections at a wrong aspect of the argument. The same will be found to apply to certain modal versions of the argument.

[5] For the idea of an exclusive reading of quantifiers, see Jaakko Hintikka, 'Identity, Variables, and Impredicative Definitions', *Journal of Symbolic Logic* **21** (1956) 225–45.

[6] I have tried to spell out the logic of the epistemic operators '*K*' and '*P*' in my book, *Knowledge and Belief*, Cornell University Press, Ithaca, N.Y., 1962. I rely on what is said there in the present paper. For some problems which one encounters in this area and for their resolution, see also the symposium on epistemic logic in the first issue of *Noûs* **1** (1967).

[7] This is the typical mode of argument employed in *Knowledge and Belief.*

[8] Cf. *Knowledge and Belief*, Section 6.8, and my paper, 'Individuals, Possible Worlds, and Epistemic Logic', *Noûs* **1** (1967) 33–62, especially 35–38.

[9] Cf. the conditions (C.P*) and (C.&) of *Knowledge and Belief.*

[10] In virtue of (10), keeping in mind that the name of some *existing* individual must be able to replace *z* in (10).

[11] For a defense of this reading, see *Knowledge and Belief*, especially pp. 131–132, and 'Individuals, Possible Worlds, and Epistemic Logic', pp. 50–53.

[12] *Knowledge and Belief*, p. 160.

[13] Castañeda has repeated this claim quite a few times. For a sample, see 'On the Logic of Self-Knowledge', *Noûs* **1** (1967) 9–21, especially p. 9.

[14] See e.g. 'Individuals, Possible Worlds, and Epistemic Logic', pp. 46–48.

III. THE SEMANTICS OF MODALITY

MODALITY AND QUANTIFICATION

Most branches of logic may be studied by means of two different (although related) methods or sets of methods which are usually called syntactical and semantical, respectively. In this paper, I shall outline some basic ideas of a semantical theory of modal logic, including quantified modal logic. Since a fuller treatment is easy to carry out on the basis of this outline, I shall omit most of the proofs.[1]

The basic notion of a semantical theory is normally the notion of truth. In so far as we are not interested in truth under some particular interpretation of logical formulae but rather in the question whether there are any interpretations which make a given set of formulae true (in short, if we are not interested in any one interpretation more than in the others), the basic concept of a semantical theory may also be chosen to be that of satisfiability.[2] If the negation of a formula p is not satisfiable, p is said to be valid.

For ordinary non-modal logic (quantification theory), the notion of satisfiability may be defined (following Carnap, with slight modifications) as follows: A set of formulae λ is satisfiable if and only if there is a state-description in which all the members of λ hold. (A formula is satisfiable if and only if its unit set is.) Now a set of formulae μ is the set of all formulae which hold in some particular state-description if and only if it satisfies the following conditions:

(C.1) If p is an atomic formula or an identity, then not both $p \in \mu$ and $\sim p \in \mu$;

(C.2) If p is an atomic formula or an identity and if all the free individual variables of p occur in the other formulae of μ, then either $p \in \mu$ or $\sim p \in \mu$;

(C.3) If p is an atomic formula or an identity, if q is like p except that a and b have been interchanged in one or more places, if $p \in \mu$, and if $a = b \in \mu$, then $q \in \mu$;

(C.4) Not $\sim (a = a) \in \mu$;

(C.5) If $(p\,\&\,q)\in\mu$, then $p\in\mu$ and $q\in\mu$;
(C.6) If $p\in\mu$ and $q\in\mu$, then $(p\,\&\,q)\in\mu$;
(C.7) If $(p \vee q)\in\mu$, then $p\in\mu$ or $q\in\mu$ (or both);
(C.8) If $p\in\mu$ or $q\in\mu$ and if all the free individual variables of $(p \vee q)$ occur in the other formulae of μ, then $(p \vee q)\in\mu$;
(C.9) If $(Ex)p\in\mu$, then $p(a/x)\in\mu$ for at least one free individual variable a;
(C.10) If $p(a/x)\in\mu$ for at least one free individual variable a, then $(Ex)p\in\mu$;
(C.11) If $(Ux)p\in\mu$ and if b occurs in at least one formula of μ, then $p(b/x)\in\mu$;
(C.12) If $p(b/x)\in\mu$ for every free individual variable b which occurs in the formulae of μ, then $(Ux)p\in\mu$.

Comments:

(1) Here the conditions (C.1)–(C.4) make sure that that part of μ which consists of atomic formulae and identities is a state-description (in a sense closely related to Carnap's). The other conditions serve to make sure, on one hand, that all the other formulae of μ hold in the state-description in question and on the other hand, that all the formulae holding in it belong to μ. These conditions are nothing but variants of the usual semantical rules for &, $\vee$, E, and U.

(2) In the conditions (C.1)–(C.12), I have made use of the following assumptions:

(i) p, q, ... are arbitrary formulae;
(ii) a, b, ... are arbitrary free individual variables (placeholders for free singular terms);
(iii) x, y, ... are arbitrary bound individual variables;
(iv) $p(a/x)$ is the formula obtained from p by replacing x everywhere by a;

(v) all the sentential connectives other than $\sim$, & and $\vee$ have been eliminated;

(vi) all the formulae we are dealing with have been brought into a 'negational miniscope form' in which negation-signs occur only where they immediately precede an atomic formula or an identity;

(vii) '$\in$' is (of course) a metalogical shorthand for 'is a member of'.

(3) If we want to exclude empty universes of discourse, it may be

accomplished by adopting the following additional condition:

(C.u) If $(Ux)p \in \mu$, then $p(a/x) \in \mu$ for at least one free individual variable a.

We may thus paraphrase our original definition and say that a set of formulae is satisfiable if and only if it can be imbedded in a set which satisfies the conditions (C.1)–(C.12). (A set which satisfies these conditions will be called an extended state-description.)

However, almost half of these conditions are redundant. I shall call a set of formulae which satisfies (C.1), (C.3), (C.4), (C.5), (C.7), (C.9), and (C.11) a *model set*, and rename these seven conditions (C.$\sim$), (C.$=$), (C.self$\neq$), (C.&), (C.$\vee$), (C.E), and (C.U), respectively. I shall show how to prove that a set λ of formulae is satisfiable if and only if it can be imbedded in a model set (i.e. if and only if there is a model set μ such that $\mu \supseteq \lambda$).

PROOF: (1) The 'only if'-part is trivial. (2) In order to prove the 'if'-part it suffices to prove the following pair of lemmata:

(A) Each model set may be imbedded in a maximal model set.
(B) Each maximal model set is an extended state-description.

For it follows from these lemmata that any set of formulae which is imbeddable in a model set is imbeddable in a maximal model set, i.e. in an extended state-description. (By the maximality of a model set μ, we mean that there is no larger model set $\nu \supset \mu$ such that each free individual variable occurring in the formulae of ν already occurs in the formulae of μ.)

Lemma (B) may be proved by verifying that if one of the conditions (C.2), (C.6), (C.8), (C.10), (C.12) is not satisfied by a model set μ, we may adjoin a new formula to μ so as to obtain a larger model set. Lemma (A) may be proved by means of Zorn's lemma in the same way as the corresponding result for sets with a property of finite character (cf. G. Birkhoff, *Lattice Theory*, New York 1948, pp. 42–3).

The result may perhaps be expressed intuitively by saying that a model set is the formal counterpart to a partial description of a possible state of affairs (of a 'possible world'). (It is, however, large enough a description to make sure that the state of affairs in question is really possible.) For it is natural to say that a set of sentences is satisfiable if and only if it can be embedded in a (partial or exhaustive) description of possible state of affairs; and this is just what we demonstrated if model sets are interpreted as such descriptions.

This idea helps us to extend the notion of satisfiability to sets of formulae which may contain modal operators. First of all, it shows us that when we are considering such formulae we cannot hope to get along by considering just one model set at a time. In discussing notions like possibility and necessity, we have to consider what happens in states of affairs different from the actual one. In our definition of satisfiability, we therefore have to consider sets of model sets. Such sets of sets we shall call *model systems*.

What conditions must model systems be subjected to? Suppose that $Mp \in \mu \in \Omega$, where Ω is a model system (and where M is to be read 'possibly'). Then clearly we have to require that p, which perhaps is not true in the state of affairs described by μ, must nevertheless be true in some other state of affairs which could have been realized instead of the one described by μ. Descriptions of such states of affairs will be called alternatives to μ. In other words, the following condition must be satisfied:

(C.M*) If $Mp \in \mu \in \Omega$, then there is in Ω at least one alternative ν to μ such that $p \in \nu$.

Suppose, again, that $Np \in \mu \in \Omega$, where Ω is a model system (and where N is to be read 'necessarily'). Then we have to require that what is said to happen necessarily happens actually:

(C.N) If $Np \in \mu$, then $p \in \mu$.

This, however, does not exhaust the 'meaning' of Np. When we say that something takes place necessarily, we say more than that it takes place actually. We say that it takes place unavoidably, that it would have taken place in all the other courses of events which could have been realized instead of the actual one. In formal terms, the following condition has to be adopted:

(C.N$^+$) If $Np \in \mu \in \Omega$, and if $\nu \in \Omega$ is an alternative to μ, then $p \in \nu$.

The conditions (C.M*), (C.N), and (C.N$^+$) suffice for a minimum of modal logic, semantically treated. The definition of satisfiability to which they give rise may be summed up as follows: A set λ of formulae is satisfiable if and only if there is a model system $\langle \Omega, R \rangle$ such that $\mu \supseteq \lambda$ for some member μ of Ω. A model system is a couple $\langle \Omega, R \rangle$ whose first member is a set of model sets each of which satisfies (C.N). The second member R is a two-place relation on Ω, called the relation of alternativeness. It is required, moreover, that the conditions (C.M*) and (C.N$^+$) are satisfied by Ω and R. In view of the equivalences $\sim Np \equiv M \sim p$ and $\sim Mp \equiv N \sim p$ the miniscope assumption can be made here, too.

The semantical system thus obtained is equivalent to a well-known syntactical (axiomatic) system of modal logic, which was first suggested by Kurt Gödel and which has been called by von Wright (in *Essay in Modal Logic*, Amsterdam 1951) the system M. The equivalence means, of course, that a formula is provable in M if and only if it is valid in our semantical system.

If we add the condition that the relation of alternativeness is transitive, we obtain a stronger system which is in the same sense equivalent to Lewis's system S4. If it is required that the relation of alternativeness is symmetric, we obtain a semantical system whose syntactical twin is obtained from M by adopting the so-called axiom (axiom schema) of Brouwer's:

$$p \supset NMp.$$

If it is required that the relation is transitive and symmetric, we obtain a system which is equivalent to Lewis's S5. By imposing a certain restriction on (C.M*) in the semantical counterparts to M and to S4, Lewis's systems S2 and S3, respectively, could similarly be given a semantical interpretation.[3]

I shall not prove these results here. Instead, I shall briefly consider the problems which arise when modality is combined with quantification (and/or identity). The use of quantifiers causes changes in the above conditions. For by means of the unmodified conditions we could prove results which are clearly counter-intuitive. For instance, we could 'prove' that the following formula is valid:

$$(1) \qquad (Ex)NP(x) \supset N(Ex)P(x).$$

Yet (1) is obviously unacceptable as a general logical principle. We may admit, for the sake of argument, that every wheel is necessarily round. Yet we would not therefore conclude that, since there happen to be wheels in existence, the existence of round objects is a necessary and unavoidable feature of the world.

A reductive argument to show the validity of (1) might run as follows: We make the counter-assumption

(2) $(Ex)NP(x) \in \mu \in \Omega$
(3) $M(Ux){\sim}P(x) \in \mu \in \Omega$

(for some model set μ and a model system Ω) which amounts to assuming that (1) is not valid, i.e. that its negation is satisfiable. This counter-assumption can then be reduced *ad absurdum*. In fact, we have from (2)

(4) $NP(a) \in \mu$

for some a by (C.E), and from (3)

(5) $(Ux){\sim}P(x) \in \nu \in \Omega$

for some alternative ν to μ by (C.M*). By (C.N$^+$) we have from (4)

(6) $P(a) \in \nu$.

By (C.U) we have from (5) and (6)

(7) ${\sim}P(a) \in \nu$,

which violates, together with (6), our condition (C.$\sim$), thus completing the reduction and thereby establishing the 'validity' of (1).

Since (1) obviously fails to be logically true, there must be a fallacious step somewhere in the argument. It is almost equally obvious how the fallacy came about. In step (7), we applied what (5) says of all the individuals existing in the possible world described by ν to the special case of a. But why do we have to think of a as existing in this possible world? Because the term occurs in a formula of ν, as attested to by (6). On what authority, then, do we assert (6)? This step is based on (C.N$^+$). However, it is obvious that the step from (4) to (6) involves something illicit. In (4) we say that a cannot fail to have the property P. From this it does not follow that a exists *and* has the property P in each alternative world. It only follows that *if* it exists in one of them, *then* it has this property. Hence step (6) is possible only on the further assumption, which is not made here, that a exists in the world described by ν.

This diagnosis can immediately be generalized. For the purpose, we may have a look at (C.E) and (C.U). They show that every free individual variable which occurs in the formulae of μ is assumed to behave in the same way as an individual term whose 'referent' (bearer) really exists in the state of affairs described by μ. The presence of a free variable in the formulae of μ, we may thus say, is the formal counterpart to the existence of its value in the state of affairs described by μ.

From this it follows that when a formula p is transferred from a model set μ to one of its alternatives – say v – we have to heed the free individual variables p contains. If one of them does not occur in the other formulae of v, then the adjunction of p to v is legitimate only if the relevant values of this free individual variable are assumed to exist not only in the state of affairs described by μ but also in that described by v. In general, this assumption cannot be made. Individuals which *de facto* exist may possibly fail to do so. For this reason, the condition (C.N$^+$) must be replaced by the following condition:

(C.N*) If $Np \in \mu \in \Omega$, if $v \in \Omega$ is an alternative to μ, and if each free individual variable of p occurs in at least one other formula of v, then $p \in v$.

It is easily seen that if (C.N$^+$) is replaced by (C.N*), (1) will not be valid any more. In fact, step (6) in the above argument fails to be justified by (C.N*), and cannot be restored by any other way in terms of this modified condition. Instead, the argument (2)–(5) virtually provides us with a counter-example to the validity of (1). In fact, $\Omega = \{\mu, v\}$, where v is an alternative to μ and where

$$\mu = \{(Ex)NP(x) \,\&\, M(Ux){\sim}P(x), (Ex)NP(x), M(Ux){\sim}P(x), NP(a), P(a)\},$$
$$v = \{(Ux){\sim}P(x), {\sim}P(b)\}$$

is a model system (in the new sense defined by means of (C.N*) instead of (C.N$^+$)) which shows that (1) is not valid.

It is possible, of course, to make the assumptions that whatever exists in a possible state of affairs exists in all the alternative states of affairs; in short, that whatever exists exists necessarily. However, it is important to realize that there is no need to make this assumption. And it is also important to realize that if it is made, we usually have to make other changes in our conditions in addition to strengthening (C.N*) to (C.N$^+$).

If the ancestral of the relation of alternativeness is called the relation of accessibility, it is seen that the following condition must be satisfied if every actually existing individual is assumed to exist necessarily:

(C.U*) If $(Ux)p \in \mu$ and if b occurs in the formulae of some model set from which μ is accessible (or in the formulae of μ), then $p(b/x) \in \mu$.

This new condition is really needed, for it may be shown that (C.N$^+$) does not entail (C.U*); nor does (C.U*) entail (C.N$^+$). However, (C.U*) and (C.N$^+$) are both consequences of the following condition, which clearly formulates exhaustively the assumption that free individual variables are transferable from a model set to its alternatives:

(C.self=*) If a occurs in at least one formula of μ and if v is an alternative to μ, then $(a=a) \in v$.

Thus we have two different systems, one of which dispenses with the assumption that all actually existing individuals exist necessarily while the other embodies this assumption. The former makes use of our original conditions except that (C.N$^+$) is replaced by (C.N*). We shall call it M. The other will be called M*. It may be obtained by adjoining to the conditions of M the additional condition (C.self=*). However, this is not the only way of formulating it. Alternatively, M* may be obtained from M by strengthening (C.N*) and (C.U) to (C.N$^+$) and (C.U*), respectively. For it may be shown that every set of formulae which is satisfiable in the resulting system is also satisfiable in M*. (The converse implication follows from what was said above.)

In order to illustrate the use of (C.self=*) for the same purpose as (C.N$^+$) (and (C.U*)), it may be pointed out that the arguments (2)–(7) above may be carried out by its means as follows: (2)–(5) as before; then

(5)* $(a=a) \in v$

from (4) by (C.self=*). Now (6) is justified by (C.N*). The last step, (7), is obtained as of old, completing the argument.

Using (C.U*) instead of (C.U) we might argue as follows: (2)–(5) as before; then (7) follows from (5) by (C.U*), and (6) is subsequently obtained by (C.N*) and not just by (C.N$^+$).

These variations of a theme illustrate how the characteristic strength of M* can be obtained in different ways.

From M and from M* we obtain two new systems by adjoining to their defining conditions the further condition (C.u). Here we may make use of either formulation of M*. If the second formulation is used, a simplification is possible in that it may be shown that there now is no need to strengthen (C.U) into (C.U*).

The system M deserves special interest because the possibility of dispensing with the stronger assumptions which characterize M* is not always perceived very clearly. In particular, these assumptions very easily steal into a syntactical (deductive) formulation of quantified modal logic.

In fact, if usual axioms and rules of inference for propositional modal logic are simply joined with the usual quantificational axioms (and, possibly, rules of inference), we normally obtain M* rather than M, without noticing how we came to make the rather strong and often illegitimate assumption that all individuals existing in one possible world always exist in all its alternatives.

In order to prevent this assumption from sneaking in, we have to qualify many of the usual axioms or rules of inference. For instance, the well-known rule of inference

$$\frac{p \supset q}{Np \supset Nq}$$

will have to be qualified by requiring that all the free individual variables of p must occur in q, too.

Similarly, even *modus ponens*

$$\frac{p, p \supset q}{q}$$

must be qualified by requiring that all the free individual variables of p must occur in q.

Within the deductive framework, the *rationale* of these qualifications is not very easy to see, as witnessed by modal logicians' failure to impose these indispensable restrictions at the early stages of quantified modal logic. The advantages of a semantical approach are illustrated by the ease of the diagnosis which led us to replace (C.N$^+$) by (C.N*).

In M* it was assumed that every free individual variable is transferable from a model set to its alternatives. We may also construct a still stronger system – it will be called M** – in which a transfer is permitted not only from a model set to its alternatives but also to arbitrary members of the same model system. M** may be obtained from M* by strengthening (C.U*) as follows:

(C.U**) If $(Ux)p \in \mu$ and if b occurs in the formulae of some model set belonging to the same model system as μ, then $p(b/x) \in \mu$.

If we are solely interested in the satisfiability of sets of formulae, there is a more economical way of obtaining M**. It may be shown that the notion of satisfiability (as we have defined it for sets of formulae) is not affected if we adjoin to the conditions of M (or M*) the following condition:

(C.acc) In every model system there is at least one model set from which all its other members are accessible.

If (C.acc) is satisfied, M** may be obtained from M* by adjoining to the conditions of M* the following 'inversion' of (C.U*):

(C.U_*) If $(Ux)p \in \mu$ and if b occurs in the formulae of some model set which is accessible from μ, then $p(b/x) \in \mu$.

Alternatively, we may adjoin to the conditions of M* a similar inversion (C.self$=_*$) of the condition (C.self$=$*).

An example of a formula which is valid in M** but not in M* is the Barcan formula

$$M(Ex)P(x) \supset (Ex)MP(x).$$

It is obvious, however, that the Barcan formula is unacceptable as a valid logical principle for most modalities. Clearly, what *can* exist need not always do so *actually*; birth control is not a logical impossibility.

If relation of alternativeness is assumed to be symmetric, the distinction between M* and M** vanishes (provided that (C.acc) is satisfied). Since the assumptions which underlie M* easily steal into the usual deductive systems of quantified modal logic, it may be expected that even the stronger illicit assumptions which underlie M** easily steal into those deductive systems in which the relation of alternativeness is tacitly assumed to be symmetric. Lewis's S5 is a case in point. This expectation is in fact fulfilled; the Barcan formula is provable in quantified S5 without any additional assumptions. (See A. N. Prior, 'Modality and Quantification in S5', *Journal of Symbolic Logic* **21** (1956) 60–2.)

It may be of some interest to see how Barcan's formula can be proved in M** in different ways. Again, the argument will be by *reductio ad absurdum*. The counter-assumption is

(8) $M(Ex)P(x) \in \mu \in \Omega$
(9) $(Ux)N{\sim}P(x) \in \mu \in \Omega$

for some suitable model system Ω. From (8) it follows that

(10) $(Ex)P(x) \in \nu \in \Omega$

for some alternative ν to μ. Furthermore,

(11) $P(a) \in \nu$

from (10) by (C.E) for some *a*. Now we can have

(12) $N{\sim}P(a) \in \mu$

from (9) and (11) by (C.U**) or by (C.U$_*$), which in turn implies, in virtue of (C.N*),

(13) ${\sim}P(a) \in \nu$,

which, together with (12), violates (C.$\sim$), thus completing the reductive argument.

If we are using (C.self$=_*$) rather than (C.U**) or (C.U$_*$), we may have instead of (12)

(11*) $(a=a) \in \mu$

from (11) by (C.self$=_*$). Then (12) follows by the plain old (C.U) and (13) by (C.N*).

If the alternativeness relation is assumed to be symmetric, μ is an alternative to ν (by symmetry), and we have (even on the sole basis of the assumptions of M*) (11)* from (11) by (C.self$=$*) etc., or for that matter (12) from (9) by (C.U*), and so on as before.

However, if no transferability assumptions are made (nor symmetry presupposed), an argument which starts from (8)–(9) virtually produces a model system Ω which shows the invalidity of the Barcan formula in M*.

In fact, all we have to do is put $\Omega = \{\mu, \nu\}$, ν alternative to μ,

$$\mu = \{M(Ex)P(x) \,\&\, (Ux)N{\sim}P(x),\ M(Ex)P(x),\ (Ux)N{\sim}P(x),\ N{\sim}P(b),\ {\sim}P(b)\},$$
$$\nu = \{(Ex)P(x),\ P(a),\ {\sim}P(b)\}.$$

Thus we have gained a great deal of flexibility. We can assume the transfer principle on which the validity of the Barcan formula is based, and even do so in a variety of ways, but we do not have to do so, as we often clearly do not want to do.

It may be asked whether it is possible in M to strengthen (C.=) to a form (we shall call it (C.=!)) in which the restriction to atomic formulae and identities is omitted. It turns out that if the condition (C.=!) is adopted, we are not moving within M (or M*) any more; we obtain a stronger system which could also be obtained by adopting the following condition:

(C.=*) If $a = b \in \mu$, if ν is an alternative to μ and if a and b occur in the formulae of ν, then $a = b \in \nu$,

or, in a still simpler form,

(C.N=!) If $a = b \in \mu$, then $N(a = b) \in \mu$.

In other words, adopting (C.=!) as distinguished from (C.=) is tantamount to assuming that all identities hold necessarily. This is very interesting, for (C.=!) is for all practical purposes identical with one form of the *principle of the substitutivity of identity*. Since it does not make much sense to assume that all identities hold necessarily, the equivalence of (C.=!) with (C.N=!) constitutes a telling argument against this version of the principle.

Nevertheless, the principle of the substitutivity of identity, even in the form considered here, is sometimes adopted (more or less tacitly) in building a syntactical (deductive) system of quantified modal logic. Small wonder, therefore, that in the resulting systems it is possible to 'prove' such highly paradoxical 'theorems' as

(14) $(a = b) \supset N(a = b)$.

Again, a quick comparison between the different ways in which (14) can be 'proved' may be instructive. We shall use $(a \neq b)$ as a shorthand for

$\sim(a=b)$. Assume

(15) $(a=b)\in\mu\in\Omega$
(16) $M(a\neq b)\in\mu\in\Omega$.

These two constitute the counter-assumption to be reduced *ad absurdum*. Ω is of course assumed to be a model system. Then we have

(17) $(a\neq b)\in\nu\in\Omega$

from (16) by (C.M*) for some alternative ν to μ. Now (C.=*) yields from (15)

(18) $(a=b)\in\nu$,

which, together with (17), contradicts (C.~), thus completing the reduction.

If we do not want to use (C.=*), we can argue as follows: (15)–(17) as before. Instead of (18), we must have

(18)* $M(a\neq a)\in\mu$

from (15)–(16) by (C.=!). This implies

(19) $(a\neq a)\in\lambda\in\Omega$

in virtue of (C.M*) for some alternative λ to μ. But (19) violates (C.self≠), thus completing the reduction in a new way.

Alternatively, using (C.N=!) we may argue as follows: (15)–(17) as before. Instead of (18), we may at first have

(18)** $N(a=b)\in\mu$

from (15) by (C.N=!). Then (18) follows from (18)** in virtue of (C.N*), once again completing the reduction.

However, none of these arguments succeeds if all the illicit assumptions are rejected, and counter-examples are then easily constructed to show the invalidity of (14).

If the relation of alternativeness is assumed to be symmetric, (C.=*) entails an 'inversion' (which may be called (C.$=_*$)) obtained from it by reversing the roles of μ and ν. In a system in which this assumption is made every pair of possibly identical individuals may be 'proved' to be actually identical. A deductive system of quantified S5 with the principle of substitutivity of identity as one of the axioms is a case in point. We have seen, however, that there is no need to assume (C. = !) as distinguished from (C. =). On the contrary, our considerations serve to show that the principle of substitutivity of identity is normally unacceptable in modal logic, at least in the particular form which has been considered here. One reason why this principle has often been adopted is modal logicians' failure to see that (C. = !) embodies much stronger and infinitely more dubious assumptions than (C. =). Once again syntactical methods are apt to hide the assumptions one is actually making. Once again, a semantical approach brings these assumptions to light and also shows how to dispense with them.

It remains to be seen, however, whether other considerations offer better support to the substitutivity principle and also whether some other version of the principle fares better than the one we have examined here.

REFERENCES

[1] I am aware that some of my observations have been anticipated in print or out of print by various logicians, including O. Becker, Haskell B. Curry, Peter Geach, Marcel Guillaume, Stig Kanger, Saul Kripke, J. C. C. McKinsey, C. A. Meredith, and perhaps still others. I shall not try to trace the exact relation of their ideas to mine.

[2] Only the different interpretations of non-logical symbols are considered here. The interpretation of logical constants (connectives and quantifiers) is assumed to be fixed.

[3] The restriction could be formulated by adding to (C.M*) the following clause: 'provided that μ is not an alternative to any other member of Ω or that there is at least one formula of the form Nq in μ'.

My attention was first drawn to the possibility of obtaining S2 and S3 along these lines by Saul Kripke.

THE MODES OF MODALITY

By the modes of modality I do not mean the changing fashions that prevail or have prevailed in the study of modal logics, although I would be tempted to comment on them, too.[1] I am referring to those modes or modifications which have given modal logic its name. In other words, I have in mind the variety of systems of modal logic and the variety of philosophically interesting interpretations which can often be given of them. The point of my paper is to recommend a specific method for the study of this variety, which to my mind constitutes a veritable embarrassment of riches. This embarrassment also affects my paper, I am afraid; the major part of it is a series of sketches for applications of my methods rather than a continuous argument.

These methods have been outlined in another paper of mine.[2] I shall begin by recapitulating their essentials. They are based on the notion of a *model set* (m.s.). A model set is a set of formulas – say μ – satisfying the following conditions:

(C. $\sim$)	If μ contains an atomic formula or an identity, it does not contain its negation.
(C. &)	If $(p \,\&\, q) \in \mu$, then $p \in \mu$ and $q \in \mu$.
(C. $\vee$)	If $(p \vee q) \in \mu$, then $p \in \mu$ or $q \in \mu$.
(C.E)	If $(Ex)p \in \mu$, then $p(a/x) \in \mu$ for at least one free individual symbol a. (Here $p(a/x)$ is the result of replacing x everywhere by a in p.)
(C.U)	If $(Ux)p \in \mu$ and if b is a free individual symbol which occurs in at least one formula of μ, then $p(b/x) \in \mu$.
(C.self $\neq$)	μ does not contain any formulas of the form $\sim(a = a)$.
(C. =)	If $p \in \mu$, $(a=b) \in \mu$, and if q is like p except for the interchange of a and b at some (or all) of their occurrences, then $q \in \mu$ provided that p and q are atomic formulas or identities.

In addition to these conditions, we need either corresponding conditions for negated formulas or else some way of reducing negated (nonatomic) formulas to unnegated ones. Both courses are very easy.

In the absence of logical constants other than sentential connectives, quantifiers, and identity, a m.s. may be thought of as a partial description of a possible state of affairs or a possible course of events ('possible world'). Although partial, these descriptions are large enough to show that the described states of affairs are really possible: in quantification theory the *satisfiability* of a set of formulas may be equated with its imbeddability in a m.s., as I have shown elsewhere.[3]

This approach may be extended to modal logic by using certain configurations of m.s.'s, called *model systems*. A model system is a set of m.s.'s on which a dyadic relation has been defined. This relation will be called the relation of alternativeness, and the sets bearing it to some given set μ will be called the *alternatives* to μ. Intuitively, they are partial descriptions of those states of affairs which could have been realized instead of the one described by μ. On the basis of this idea, it is seen at once that the following conditions have to be satisfied by each model system Ω and by its alternativeness relation:[4]

(C.N) If $Np \in \mu \in \Omega$, then $p \in \mu$.

(C.M*) If $Mp \in \mu \in \Omega$, then there is in Ω at least one alternative to μ which contains p.

(C.N$^+$) If $Np \in \mu \in \Omega$ and if ν is an alternative to μ in Ω, then $p \in \nu$.

Of course, we must also set up similar conditions for negated formulas, or else reduce them to unnegated ones. Both these things are easy to accomplish.

The content of (C.M*) and (C.N$^+$), respectively, may be expressed fairly accurately by saying that whatever is possible must be true in some alternative world and that whatever is necessary must be true in all the alternative worlds. What we have here is therefore a slightly modified version of the traditional idea that possibility equals truth in some 'possible world' while necessity equals truth in all 'possible worlds'. Apart from using the notion of a m.s. as an explication of the notion of a description of a possible world, our only departure from the traditional idea lies in rejecting the presupposition that all 'possible worlds' are on a par. We have assumed that not every possible world (say P) is really an alternative to a given possible world (say Q) in the sense that P could have been realized instead of Q. We have assumed, moreover, that only these genuine alternatives really count. Each statement has to be thought of as having been made in some 'possible world'; and nothing can be said to be possible in such a world which would not have been true in some world realizable in its stead. Hence the use of the alternativeness relation and the consequent appearance of the phrases 'some *alternative* possible world' and 'all *alternative* possible worlds' where you probably expected the simpler phrases 'some possible world' and 'all possible worlds', respectively.

Here we already have a theory of modal logic in a nutshell. The *satisfiability* of a set of formulas may be defined as its imbeddability in a member of a model system. (This, it is seen, is a natural generalization of the corresponding definition for quantification theory.) Even more generally, the satisfiability of an arbitrary set of sets of formulas which has an arbitrary dyadic relation (we shall call that, too, an 'alternativeness relation') defined on it may be defined as the possibility of mapping it homomorphically into a model system so that each element is included in its image (both are of course sets). Other notions, for instance those of validity, inconsistency, and logical consequence, may be defined in terms of satisfiability in the usual way.

Because of the prominence of the notion of satisfiability in this approach it may perhaps be called *semantical*. It is not very difficult to obtain a kind of *syntactical* treatment of modality, too, from the same basic ideas. Given a set of formulas, how can we hope to show that it is satisfiable? An answer is immediately suggested by the form of the conditions which define a m.s. and a model system. With the sole exception of (C.$\sim$) and (C.self$\neq$), they are all closure conditions or very much like closure conditions. (The exceptional conditions may be considered as a kind of consistency conditions.) In other words, whenever a set Σ of sets (on which an alternativeness relation has been defined) violates one of the conditions which define a model system (other than (C.$\sim$)), this violation may be removed by adjoining a new formula to one of the members of Σ or (in the case of (C.M*)) by adjoining a new member of the form $\{F\}$ to Σ to serve as an alternative to one of the old members. It can be shown without difficulty that adjunctions of this kind preserve the satisfiability of a satisfiable Σ. (In the case of (C.$\vee$) at least one of the two adjunctions which may satisfy it preserves its satisfiability.) One natural way of trying to see whether Σ is satisfiable is therefore to carry out successive adjunctions of this kind so as to try to build a model system which would show that Σ is satisfiable. If all the alternative ways of trying to do so end up in a violation of (C.$\sim$) or of (C.self$\neq$), then we know that Σ is *not* satisfiable. It is also possible to show that every proof of a formula in a suitable system of modal logic can be thought of as such an abortive attempt to build a counterexample to it, i.e. to build a model system which would show that its negation is satisfiable.[5]

It is not immediately obvious that every inconsistent set of formulas can be shown to be inconsistent by means of an abortive model system construction of this kind. It can be proved, however, that this is always possible if we conduct the attempt in a suitable way (essentially, if we do not forget any possibility of adjunction for good). This result, which will not be proved here, constitutes an interesting *completeness theorem*.

All this gives us but one system of modal logic, albeit one of the most natural systems. If we look away from its quantificational aspects, it turns out to be a semantical counterpart to that deductive system of modal logic which is perhaps best known as von Wright's system M, presented in his *Essay in Modal Logic*, North-Holland Publ. Co., Amsterdam, 1951, although it was suggested much earlier by Gödel.[6] (We shall call the corresponding semantical system 'system M', too.) The natural way in which we have arrived at this system constitutes, in my opinion, a very strong argument for its interest and importance. There are other interesting systems of modal logic, however, and we must therefore be able to modify the defining conditions of our system M so as to be able to cope with them.

There are several widely different possibilities of modification. Some of the systems which result from these modifications are semantical counterparts to well-known deductive systems; others are important for the interpretation of modal logics. Among the relevant possibilities of modification there are the following:

(1) The alternativeness relation may be assumed to have properties additional to those imposed on it by the above conditions. Conversely, it may lack some of these. For instance, it is seen that (C.N) is, in the presence of (C.N^{+}) or some similar condition, tantamount to the requirement that the alternativeness relation be reflexive. This requirement

may be given up. Then it is often advisable to adopt a weaker condition which ensures that the necessity-operator N is at least as strong as the possibility-operator M, e.g. as follows:

(C.n*) If $Np \in \mu \in \Omega$, then there is in Ω at least one alternative to μ which contains p.

Conversely, we may require that the alternativeness relation be not only reflexive but also symmetric or transitive or both. Thus transitivity gives rise to a semantical counterpart to Lewis's S4, and the combined requirement of symmetry and transitivity (plus reflexivity, of course) to a counterpart to S5. These modifications have been briefly commented on elsewhere.[7]

A remark on the semantical counterpart of S5 may have some philosophical interest. A transitive and symmetric relation is sometimes known as an equivalence relation: it effects a partition of its field into equivalence classes in such a way that two different members of the same class always bear this relation to each other while members of different classes never bear it to each other. In the case of our semantical version of S5, we may for certain purposes require that there is but one such equivalence class. (For instance, this requirement does not affect the satisfiability of any sets of formulas.) And if we do so, the situation will begin to seem rather familiar: every m.s. is an alternative to every other m.s. This is indeed the situation presupposed by the traditional identification of possibility with truth in some possible world and of necessity with truth in every possible world. The fact that the traditional idea thus yields only a rather special kind of modal system perhaps serves to explain why the traditional idea was not very fruitful in the theory of modal logic for a long time, and motivates our departure from the tradition.

(2) We may modify the assumptions which pertain to quantification. Modifications of this kind are sometimes independent of the presence of modal notions. But even so, they are made desirable by the interplay of modality and quantification. The most fundamental modification of this kind is the elimination of what I have called *existential presuppositions*.[8] They are presuppositions to the effect that all our singular terms refer to some actually existing individual, i.e. that empty singular terms are excluded from the discussion. In an uninterpreted system, this of course

means that free individual symbols must not behave like empty singular terms.

Presuppositions of this kind are made, usually tacitly, in all the traditional systems of quantification theory. In terms of our model set technique, they are especially easy to eliminate. All we have to do is to modify (C.E) and (C.U) as follows:

(C.E_0) If $(Ex)p \in \mu$, then $p(a/x) \in \mu$ and $(Ex)(x=a) \in \mu$ for at least one free individual symbol *a*.

(C.U_0) If $(Ux)p \in \mu$ and $(Ey)(y=b) \in \mu$ (or $(Ey)(b=y) \in \mu$), then $p(b/x) \in \mu$.

Explanation: The formula $(Ex)\,(x=a)$ naturally serves as the formalization of the phrase '*a* exists'. (Cf. Quine's dictum 'to be is to be a value of a bound variable' which for our purposes might be expanded to read 'to be is to be identical with one of the values of a bound variable'.) Accordingly, the new condition (C.U_0) says that whatever is true of all actually existing individuals is true of the individual referred to by *b* *provided that such an individual really exists.* If empty singular terms are admitted to our systems, the italicized provided-clause is obviously needed. Similarly, the additional force of (C.E_0) over and above that of (C.E) is seen to lie in the requirement that the term *a*, which serves to represent one of those individuals which are being claimed to exist by $(Ex)p$, really refers to some actually existing individual (or, if we are dealing with an uninterpreted system, behaves as if it did). This strengthening is made necessary by the admission of empty singular terms.

The system obtained by eliminating the existential presuppositions is weaker than our original system M. All those inferences, exemplified by the existential generalization, which turn on the exclusion of empty singular terms are now invalid. They are restored, however, by means of contingent extra premises of the form $(Ey)(y=b)$. For instance, although the formula $p(a/x) \supset (Ex)p$ is not valid any more, the closely related formula $(p(a/x) \,\&\, (Ex)(x=a)) \supset (Ex)p$ is valid.

(3) This does not yet solve the much-discussed problems of combining modality with quantification. There is a way out of these difficulties, however, for which I have argued (in a particular case) in another context.[9] Here we shall consider only sentences with no iterated modalities. In this case, our way out is completely analogous to the elimination

of existential presuppositions which we just accomplished. All we have to do is to give the formula $(Ex)N(x=a)$ a role similar to the role which the formula $(Ex)(x=a)$ plays in the elimination of existential presuppositions: Whenever there are occurrences of x within the scope of modal operators in p we modify (C.E) (or ($C.E_0$)) by making the presence of $(Ex)\,p$ in μ imply the presence of $(Ex)N(x=a)$ in μ; and we modify (C.U) (or ($C.U_0$)) by making its applicability conditional on the presence of a formula of the form $(Ey)N(y=b)$ or $(Ey)N(b=y)$ in μ.

These modifications effect a further weakening of our system. The critical inferences whose feasibility was at issue will now depend on contingent premises of the form $(Ex)N(x=a)$ or $(Ex)N(a=x)$.

It may be argued that these modifications give us a way of meeting the objections of those logicians who have doubted the feasibility (or the advisability) of quantifying into modal contexts.[10] The gist of these objections has been, if I have diagnosed them correctly, that a genuine substitution-value of a bound individual variable must be a singular term which really specifies a well-defined individual, and that an ordinary singular term may very well fail to do so in a modal context. For instance, from

(i) the number of planets is nine but it is possible that it should be larger than ten

(which may be assumed to be true for the sake of argument) we cannot infer

(ii) $(Ex)\,(x=9$ & it is possible that $x>10)$,

for in so far as (ii) makes sense, it appears to be obviously false.

The reason for this failure is connected with the fact that the singular term 'the number of planets' in (i) does not specify any well-defined number such as is asserted to exist in (ii). (Is this number perhaps 9? But 9 cannot possibly be larger than 10. If it is not 9, what is it?) Yet the step from (i) to (ii) is justified by our unmodified conditions (C.E) and (C.U). Hence something is wrong with our system, and it is easily seen that the elimination of existential presuppositions does not help us.

It seems to me that these objections are entirely valid, and that they must be met by anybody who presumes to work out a system of quantified modal logic. A way of meeting them is perhaps seen by asking: Why do some terms fail in modal contexts to have the kind of unique reference which is a prerequisite for being a substitution-value of a bound variable? An answer is implicit in our method of dealing with modal logic. Why does the term 'the number of planets' in (i) fail to specify a well-defined individual? Obviously because in the different states of affairs which we consider possible when we assert (i) it will refer to different numbers. (In the actual state of affairs it refers to 9, but we are also implicitly considering other states of affairs in which it refers to larger numbers.) This at once suggests an answer to the question as to when a singular term (say a) really specifies a well-defined individual and therefore qualifies as an admissible substitution-value of the bound variables. It does so if and only if it refers to one and the same individual not only in the actual world (or, more generally, in whatever possible world we are considering) but also in all the alternative worlds which could have been realized instead of it; in other words, if and only if there is an individual to which it refers in all the alternative worlds as well. But referring to it in all these alternatives means referring to it *necessarily*. Hence $(Ex)N(x=a)$ formulates an obvious necessary and sufficient condition for the term a to refer to a well-defined individual in the sense the critics of quantified modal logic seem to have been driving at, exactly as I suggested.

Other modal logicians have preferred to let all the free individual symbols of a logical system be admissible substitution-values of bound individual variables. Then they have had to restrict the class of singular terms which in an interpretation may be substituted for free individual symbols to those which have the desired kind of unique reference. This procedure is certainly feasible, but it seems to me to restrict the applicability of our logical system far too much. These limitations are especially heavy in areas where even proper names might fail to have the required sort of well-defined reference and hence might not qualify as substitution-values of free individual symbols. This seems to happen in epistemic logic. In *Knowledge and Belief*[11] I argued that in epistemic logic the well-defined reference with which we are here concerned is tantamount to *known* reference. If so, proper names may certainly fail to have it, for one may very well fail to know to whom a certain proper name refers. And if proper names fail us, what does not?

(4) We may also attempt modifications in an entirely different direction. We may, or we may not, assume that individuals existing in one state

of affairs always exist in the alternative states of affairs. Conversely, we may, or we may not, assume that individuals existing in one of the alternatives to a given state of affairs always exist in this given state itself. In a system *without* existential presuppositions these assumptions may be formalized very simply by assuming the transferability of formulas of the form $(Ex)\,(x=a)$ or $(Ex)\,(a=x)$ from a model set to its alternatives or *vice versa*. In systems *with* existential presuppositions the situation is more complicated. The simplest and most flexible system is the one in which no transferability assumptions are made. In order to reach such a system, we must in fact modify (C.N$^+$) so as to make its applicability conditional on the occurrence of each free individual symbol of p in at least one formula of μ. The rationale of this modification is straightforward: From (C.U) it is seen that in systems with existential presuppositions the mere presence of a free individual symbol in the members of a m.s. presupposes that it refers to an actually existing individual. In order not to assume that individuals always transfer from one possible world to its alternatives, we must therefore avoid assuming that free individual symbols may be transferred from a m.s. to its alternatives.

The ways in which such transferability assumptions may be formulated in systems with existential presuppositions have been discussed briefly in another paper of mine.[12]

(5) Thus far, we have been concerned with ways of obtaining new systems of modal logic. There are other methods of variation, however, viz. methods of formulating the assumptions of any given system in different ways some of which may often be more useful or more illuminating for certain particular purposes than others. An especially useful strategy in this connection is to replace 'global' conditions pertaining to the alternativeness relation at large by 'local' ones governing the relation of a m.s. to its alternatives. A typical example of global conditions is the requirement of transitivity. It is not very difficult to show[13] that this condition can be replaced (unless further conditions are present in addition to those of M) by the following local condition:

(C.NN$^+$) If $Np \in \mu \in \Omega$ and if ν is an alternative to μ in Ω, then $Np \in \nu$.

If this condition is fulfilled and if the other conditions are those of M, the effect of the requirement of symmetry may be obtained by adding the following condition which is again of the 'local' type:

(C.NN$_+$) If $Np \in \nu \in \Omega$ and if ν is an alternative to μ in Ω, then $Np \in \mu$.

It is fairly obvious, in fact, that this condition would give us the effects of symmetry. We have assumed that the only other conditions regulating the relation of all alternatives are (C.NN$^+$) and (C.N$^+$). The latter, however, is a consequence of the former and (C.N). Hence the only relevant condition is (C.NN$^+$), and (C.NN$_+$) is its mirror image.

We may also combine two or more conditions into one. For instance, part of the combined power of the conditions (C.M*) and (C.N$^+$) can obviously be obtained by using the following condition:

(C.M&N$^+$) If $Mq \in \mu \in \Omega$, $Np_1 \in \mu$, $Np_2 \in \mu, \ldots, Np_k \in \mu$, then there is an alternative to μ in Ω which contains all the formulas $q, p_1, p_2, \ldots, p_k$.

In a sense, even the whole power of (C.N$^+$) is obtained by means of this rule, viz. in the sense that every set of formulas which is satisfiable remains satisfiable when (C.M&N$^+$) replaces (C.N$^+$), and *vice versa*. This is not obvious. In case some m.s. μ contains an infinity of formulas of the form Np, it follows from (C.N$^+$) that all its alternatives likewise contain an infinity of formulas. Our new condition (C.M&N$^+$) only guarantees that there are alternatives to μ which contain any given finite subset of the infinity of formulas which (C.N$^+$) squeezes into each alternative of μ. That the weaker-looking condition (C.M&N$^+$) nevertheless suffices is suggested, although not quite proved yet, by the completeness theorem mentioned above. It says that every set of formulas which is not satisfiable may be shown to be so by trying to construct a model system for it; if we proceed in a suitable way, all the possible ways of trying to construct one end up in a violation of (C.$\sim$) after some finite number of steps. Now because of this finitude only a finite number of applications of (C.N$^+$) are needed in the argument. And any such finite number of applications of (C.N)$^+$ can be shown to be replaceable by an application of (C.M&N$^+$).

If the requirement of transitivity is added to our basic system M, the condition (C.M&N$^+$) must be replaced by the following condition:

(C.M&NN$^+$) If $Mq \in \mu \in \Omega$, $Np_1 \in \mu$, $Np_2 \in \mu, \ldots$, $Np_k \in \mu$, then there is in Ω at least one alternative to μ which contains all the formulas q, Np_1, $Np_2, \ldots$, Np_k.

Such variations of our original conditions might be produced almost indefinitely. They are sometimes of technical interest. More importantly, they often help us in the interpretation of the different systems of modal logic, and in finding systems of modal logic to formulate the structure of the various philosophically interesting notions to the analysis of which we may wish to apply our industry. I shall give you a few instances of such formulations.

If we are doing tense-logic, more explicitly, if we read '*M*' as 'it is or will be the case that', then our 'possible worlds' may be given a clear-cut meaning: they are the states of the world at the different moments of the future. Each m.s. is a set of true statements that can all be made at one and the same moment of (future) time. A m.s. is an alternative to another if and only if the moment of time thus correlated with the former is later than that correlated with the latter. Then it is readily seen that a formulation of this tense-logic is obtained by taking our basic system M and adding to its defining conditions the requirement that the alternativeness relation must effect a *linear* ordering.[14]

More accurately, what we obtain in this way is a tense-logic which goes together with classical physics. If we do not want to tie our logic to old-fashioned physics, we are undoubtedly wiser if we interpret each m.s. as a set of true statements that can all be made in one and the same world-point (point-instant) and read '*M*' as 'it will be the case somewhere that'. Then we can no longer require that the alternativeness relation (in this case it could perhaps be more appropriately termed 'futurity relation') effect a linear ordering. At the relativistic best, we seem to have only transitivity. Hence S4 is perhaps not so inappropriate as a system of tense-logic after all.

We can also see which system of modal logic recommends itself as the formalization of *logical* possibility and *logical* necessity. It seems to me obvious that whatever is logically necessary here and now must also be logically necessary in all the logically possible states of affairs that could have been realized instead of the actual one. (It is logically possible that this Colloquium had not been held; but no logical truths would undoubtedly have been destroyed as a consequence.) But this is exactly what (C.NN$^+$) says. Conversely, it also seems fairly clear that no new logical necessities can come about as a result of the realization of any logical possibility. In short, it seems to me that whatever is logically necessary in one logically possible world must also be logically necessary in others. But this presupposes that (C.NN$_+$) is also satisfied. Together with (C.NN$^+$), this condition imposes the structure of Lewis's S5 on our modal logic. The system S5, then, seems to be the best formalization of our logic of *logical* necessity and *logical* possibility.

An important qualification is in order here, however. In the argument I just gave no reference was made to any way of actually finding out which sentences are logically necessary or logically possible. This is essential for the correctness of the argument. In fact, we obtain entirely different results if we consider, not logical truths 'by themselves', but such logical truths as can be actually proved by means of some definite class of arguments, e.g. in some given deductive system. Then 'Np' will say that p can be proved by means of the arguments in question, and 'Mp' will say that p cannot be disproved by their means. But if so, a formula like

(iii) $$Mp \supset NMp$$

is not likely to be valid, for what it would amount to is to say that if p cannot be disproved by means of a certain class of arguments, then it can be proved by means of those very arguments that we cannot disprove it. And this is in most cases false. Nevertheless, (iii) is valid in S5, as you can verify without any difficulty. Hence S5 cannot really serve as a formalization of *provable* logical truth. It may even be doubted whether all the laws of S4 are valid in this case, as witnessed by Gödel's comments on the formula $N(Np \supset p)$.[15] If so, Halldén's conclusions[16] may have to be modified.

Further interpretations are indicated in earlier works of mine. In one of them, I discussed along these lines the conditions which our normative notions must satisfy. (A largely new discussion of this subject is given in 'Deontic Logic and Its Philosophical Morals', the present volume pp. 184–214.) In another work,[17] I examined our principal epistemic notions, viz. those of knowledge and belief, in the same respect. The former of these is simple enough to be discussed here. Suppose someone makes a number of statements on one and the same occasion, including the following: 'it is possible, for all that I know, that q'; 'I know that p_1', 'I know that p_2',..., 'I know that p_k'. When is he consistent? It seems clear that if it is really possible, for all that he knows, that q should be the case, then it must be possible for q to turn out to be the case while everything he says he knows is also true. On the interpretation 'N' = 'I know that', 'M' = 'it is possible, for all that I know, that', this is exactly what (C.M&N$^+$) says. But this is not enough. If our man really knows what he claims he knows, i.e. knows it in the sense in which knowledge is contrasted to true opinion, then it must be possible for q to turn out to be the case while he continues to know everything he claims to know. In other words, the realization of whatever is possible, for all that he knows, must not force him to give up any of his claims to knowledge, if he is to be as much as self-consistent. But this is exactly what the condition (C.M&NN$^+$) requires. And the satisfaction of this condition means, we have seen, that the logic of knowledge is at least as strong as Lewis's S4. (It is, as far as I can see, exactly tantamount to S4, provided that we forget a number of qualifications which I have considered in some detail in *Knowledge and Belief* and also forget the fact that epistemic notions are normally relative to a person.)

On the other hand, if our man only aspires to true opinion, the situation is different. There is no inconsistency in his giving up one of his opinions when something which may be true according to his (true) opinion turns out to be the case. In other words, the logic of true belief is not that of S4, although the logic of 'real' knowledge is. Hence our approach suggests an interesting, albeit partial, answer to the time-honoured question concerning the difference between 'genuine' knowledge and 'mere' true opinion. If I am right, the two notions even have different logics.

Since this is a very interesting point, it may be worth elaborating further. With more justification than perhaps meets the eye, the import of (C.M&N$^+$) may be said to consist in the requirement that if μ is satisfiable and if $Mq \in \mu$, $Np_1 \in \mu$, $Np_2 \in \mu, \ldots, Np_k \in \mu$, then the set $\{q, p_1, p_2, \ldots, p_k\}$ is also satisfiable. The latter set is satisfiable if and only if the implication

$$q \supset (\sim p_1 \vee \sim p_2 \vee \cdots \vee \sim p_k)$$

is *not* valid (logically true). When used in connection with the epistemic notions, the import of (C.M&N$^+$) may thus be said to consist in the requirement that nothing that is compatible with what somebody knows may qualify as a destructive objection to *what he says he knows.*

Likewise, the import of the stronger condition (C.M&NN$^+$) in epistemic contexts may be said to consist in the requirement that nothing that is compatible with what somebody knows may amount to a destructive objection to *his claim to know* what he says he knows. For its gist may be shown to lie in the requirement that the satisfiability of μ (of the kind mentioned in (C.M&NN$^+$)) entails that the implication

$$q \supset (\sim Np_1 \vee \sim Np_2 \vee \cdots \vee \sim Np_k)$$

is not valid. From this it is seen that if we are dealing with genuine knowledge and not just with true opinion, (C.M&NN$^+$) has to be fulfilled. For somebody's claim to knowledge in this sense of the word can be criticized not only by showing that the facts are not as he claims to know they are but also by showing that *he does not really know* (is not in the position or in the condition to know it, or whatnot) that they are as they are. By the same token, mere true opinion does *not* satisfy (C.M& NN$^+$), although it satisfies (C.M&N$^+$). (Of course, much of what passes as knowledge in ordinary discourse is, in the sense of our artificially precise distinction, mere true opinion.)

REFERENCES

[1] In his paper, 'The Philosophical Significance of Modal Logic', *Mind*, n.s., **69** (1960) 466–485, Gustav Bergmann has presented interesting criticisms of earlier treatments of modal logic. As far as the situation at the time of the writing of Bergmann's article is concerned, I agree with what I take to be his main point, viz. the failure of all the earlier deductive systems of modal logic to be based on a satisfactory semantical (or, if you prefer, combinatorial) characterization of validity (logical truth). It seems to me, however, that in this respect the situation is now radically different. I am convinced that in such papers as Stig Kanger, *Provability in Logic* (Stockholm Studies in Philosophy, vol. I), Stockholm 1957; Stig Kanger, 'The Morning Star Paradox', *Theoria* **23** (1957) 1–11; Stig Kanger, 'A Note on Quantification and Modalities', *ibid.* 133–134; Stig Kanger, 'On the Characterization of Modalities', *ibid.* 152–155; Saul Kripke, 'A Completeness Theorem in Modal Logic', *The Journal of Symbolic Logic* **24** (1959) 1–14; Saul Kripke, 'Semantical Considerations on Modal Logic' (*Proceedings of a Colloquium on Modal and Many-Valued Logics, Helsinki, 23–26 August, 1962*), *Acta Philosophica Fennica* **16** (1963) 83–94; Saul Kripke, 'Semantical Analysis of Modal Logic: I, Normal Modal Propositional Calculi', *Zeitschrift für mathematische Logik und Grundlagen der Mathematik* **9** (1963) 67–96; Saul Kripke, 'Semantical Analysis of Modal Logic: II, Non-Normal Modal Propositional Calculi' in *The Theory of Models*

(Proceedings of the 1963 International Symposium at Berkeley, ed. by J. W. Addison, L. Henkin, and A. Tarski), Amsterdam 1965, pp. 206–220; Saul Kripke, 'The Undecidability of Monadic Modal Quantification Theory', *Zeitschrift für mathematische Logik und Grundlagen der Mathematik* **8** (1962) 113–116; Richard Montague, 'Logical Necessity Physical Necessity, Ethics, and Quantifiers', *Inquiry* **3** (1960) 259–269; Richard Montague and Donald Kalish, 'That', *Philosophical Studies* **10** (1959) 54–61; my papers in this volume, and in still others we have the beginnings of the kind of foundation for modal logics which Bergmann missed.

2 Jaakko Hintikka, 'Modality and Quantification', *Theoria* **27** (1961) 119–128. Reprinted in the present volume, pp. 57–70.

3 Jaakko Hintikka, 'Form and Content in Quantification Theory', *Acta Philosophica Fennica* **8** (1955) 7–55; and Jaakko Hintikka, 'Notes on Quantification Theory', *Societas Scientiarum Fennica, Commentationes phys.-math.* **17** (1955), no. 12.

4 As I shall proceed to point out, however, there are many ways of modifying these conditions. A modification of ($C.N^+$) is sometimes necessary; see Jaakko Hintikka, 'Modality and Quantification', *Theoria* **27** (1961) 124–125; reprinted in the present volume, pp. 57–70.

5 Given a set of formulas for which we are trying to construct a model system, it does not usually suffice to consider just one m.s. together with suitable alternatives to it. Normally, we have to consider alternatives to these alternatives, and so on. In fact, for every finite integer k we can easily find a set of formulas (and even a single formula) such that more than k m.s.'s have to be considered in order to show that it is satisfiable (or that it is not satisfiable). Since each m.s. may be thought of as describing a 'possible world' in which sentences are true or false, it is to be expected that no set of truth-table-like matrices with only a finite number of truth-values serves to define such concepts as satisfiability and validity for our system. Small wonder, therefore, that the corresponding result for provability has in fact been proved for a number of Lewis's modal systems by James Dugundji in his 'Note on a Property of Matrices for Lewis and Langford's Calculi of Propositions', *The Journal of Symbolic Logic* **5** (1940) 150–151 (cf. also Kurt Gödel, 'Zum intuitionistischen Aussagenkalkül' in *Akademie der Wissenschaften in Wien, Mathematisch-naturwissenschaftliche Klasse* **69** (1932), 65–66; published also in *Ergebnisse eines mathematischen Kolloquiums*, vol. IV (for 1931–32, published in 1933), p. 40). I fail to see why his results should show that a satisfactory semantical theory of modal logic is impossible, as has been alleged. On the contrary, it would seem quite unnatural if we could fix once and for all a finite upper limit to the number of the possible worlds we have to consider in a semantical theory of modal logic.

6 Kurt Gödel, 'Eine Interpretation des intuitionistischen Aussagenkalküls' in *Ergebnisse eines mathematischen Kolloquiums*, vol. IV (for 1931–32, published 1933), pp. 39–40.

7 Jaakko Hintikka, 'Modality and Quantification', *Theoria* **27** (1961) 119–128. Reprinted in the present volume, pp. 57–70.

8 See Jaakko Hintikka, 'Existential Presuppositions and Existential Commitments', *The Journal of Philosophy* **56** (1959) 125–137, and Jaakko Hintikka, *Knowledge and Belief: An Introduction to the Logic of the Two Notions*, Ithaca, N.Y. 1962, pp. 129–131.

9 Jaakko Hintikka, *Knowledge and Belief: An Introduction to the Logic of the Two Notions*, Ithaca, N.Y., 1962, pp. 138–158.

10 See e.g. W. V. Quine, *From a Logical Point of View: 9 Logico-Philosophical Essays*, 2nd ed., revised, Cambridge, Mass., 1961; W. V. Quine, 'Quantifiers and Propositional Attitudes', *The Journal of Philosophy* **53** (1956) 177–187; W. V. Quine, *Word and Object*,

New York and London 1960; and cf. Dagfinn Føllesdal, *Referential Opacity and Modal Logic*, unpublished doctoral dissertation at Harvard University 1961, deposited in Widener Library, Harvard University.

[11] Jaakko Hintikka, *Knowledge and Belief: An Introduction to the Logic of the Two Notions*, Ithaca, N.Y., 1962, pp. 148–154.

[12] Jaakko Hintikka, 'Modality and Quantification', *Theoria* **27** (1961) 119–128. Reprinted in the present volume, pp. 57–70.

[13] Cf. Jaakko Hintikka, *Knowledge and Belief: An Introduction to the Logic of the Two Notions*, Ithaca, N.Y., 1962, pp. 46–47.

[14] Hence S4 is not yet a satisfactory system of this kind of tense-logic, for in S4 the alternativeness relation need not yet be linear; it is only required to be transitive. Cf. A. N. Prior, *Time and Modality*, Oxford 1957, and Jaakko Hintikka, 'Review of Prior's *Time and Modality*', *The Philosophical Review* **67** (1958) 401–404.

[15] Kurt Gödel, 'Eine Interpretation des intuitionistischen Aussagenkalküls', in *Ergebnisse eines mathematischen Kolloquiums*, vol. IV (for 1931–32, published 1933), pp. 39–40.

[16] Sören Halldén, 'A Pragmatic Approach to Modal Logic' in *Filosofiska studier tillägnade Konrad Marc-Wogau 4 april 1962* (ed. by Ann-Mari Henschen-Dahlquist and Ingemar Hedenius), Uppsala 1962, pp. 82–94.

[17] Jaakko Hintikka, *Knowledge and Belief: An Introduction to the Logic of the Two Notions*, Ithaca, N.Y., 1962, pp. 16–22, 23–29.

SEMANTICS FOR PROPOSITIONAL ATTITUDES

I. THE CONTRAST BETWEEN THE THEORY OF REFERENCE AND THE THEORY OF MEANING IS SPURIOUS

In the philosophy of logic a distinction is often made between the *theory of reference* and the *theory of meaning*.[1] In this paper I shall suggest (*inter alia*) that this distinction, though not without substance, is profoundly misleading. The theory of reference is, I shall argue, the theory of meaning for certain simple types of language. The only entities needed in the so-called theory of meaning are, in many interesting cases and perhaps even in all cases, merely what is required in order for the expressions of our language to be able to refer in certain more complicated situations. Instead of the theory of reference and the theory of meaning we perhaps ought to speak in some cases of the theory of simple and of multiple reference, respectively. Quine has regretted that the term 'semantics', which etymologically ought to refer to the theory of meaning, has come to mean the theory of reference.[1] I submit that this usage is happier than Quine thinks, and that large parts of the theory of meaning in reality are – or ought to be – but semantical theories for notions transcending the range of certain elementary types of concepts.

It seems to me in fact that the usual reasons for distinguishing between meaning and reference are seriously mistaken. Frequently, they are formulated in terms of a first-order (i.e., quantificational) language. In such a language, it is said, knowing the mere references of individual constants, or knowing the extensions of predicates, cannot suffice to specify their meanings because the references of two individual constants or the extensions of two predicate constants 'obviously' can coincide without there being any identity of meaning.[2] Hence, it is often concluded, the theory of reference for first-order languages will have to be supplemented by a theory of the 'meanings' of the expressions of these languages.

The line of argument is not without solid intuitive foundation, but its implications are different from what they are usually taken to be. This whole concept of meaning (as distinguished from reference) is very unclear and usually hard to fathom. However it is understood, it seems to me in any case completely hopeless to try to divorce the idea of the meaning of a sentence from the idea of the *information* that the sentence can convey to a hearer or reader, should someone truthfully address it to him.[3] Now what is this information? Clearly it is just information to the effect that the sentence is true, that the world is such as to meet the truth-conditions of the sentence.

Now in the case of a first-order language these truth-conditions cannot be divested from the references of singular terms and from the extensions of its predicates. In fact, these references and extensions are precisely what the truth-conditions of quantified sentences turn on. The truth-value of a sentence is a function of the references (extensions) of the terms it contains, not of their 'meanings'. Thus it follows from the above principles that a theory of reference is for genuine first-order languages the basis of a theory of meaning. Recently, a similar conclusion has in effect been persuasively argued for (from entirely different premises and in an entirely different way) by Donald Davidson.[4] The references, not the alleged meanings, of our primitive terms are thus what determine the meanings (in the sense explained) of first-order sentences. Hence the introduction of the 'meanings' of singular terms and predicates is strictly useless: In any theory of meaning which serves to explain the information which first-order sentences convey, these 'meanings' are bound to be completely idle.

What happens, then, to our intuitions concerning the allegedly obvious difference between reference and meaning in first-order languages? If these intuitions are sound, and if the above remarks are to the point, then the only reasonable conclusion is that our intuitions do not really pertain to first-order discourse. The 'ordinary language' which we think of when we assert the obviousness of the distinction cannot be reduced to the canonical form of an applied first-order language without violating these intuitions. How these other languages enable us to appreciate the real (but frequently misunderstood) force of the apparently obvious difference between reference and meaning I shall indicate later (see Section VI *infra*).

II. FIRST-ORDER LANGUAGES

I conclude that the traditional theory of reference, suitably extended and developed, is all we need for a full-scale theory of meaning in the case of an applied first-order language. All that is needed to grasp the information that a sentence of such a language yields is given by the rules that determine the references of its terms, in the usual sense of the word. For the purposes of first-order languages, to specify the meaning of a singular term is therefore nearly tantamount to specifying its reference, and to specify the meaning of a predicate is for all practical purposes to specify its extension. As long as we can restrict ourselves to first-order discourse, the theory of truth and satisfaction will therefore be the central part of the theory of meaning.

A partial exception to this statement seems to be the theory of so-called 'meaning postulates' or 'semantical rules' which are supposed to catch non-logical synonymies.[5] However, I would argue that whatever non-logical identities of meaning there might be in our discourse ought to be spelled out, not in terms of definitions of terms, but by developing a satisfactory semantical theory for the terms which create these synonymies. In those cases in which meaning postulates are needed, this enterprise no longer belongs to the theory of first-order logic.

In more precise terms, one may thus say that to understand a sentence of first-order logic is to know its interpretation in the actual world. To know this is to know the interpretation function ϕ. This can be characterized as a function which does the following things:

(1.1) For each individual constant a of our first-order language, $\phi(a)$ is a member of the domain of individuals **I**.

The domain of individuals **I** is of course to be thought of as the totality of objects which our language speaks of.

(1.2) For each constant predicate Q (say of n terms), $\phi(Q)$ is a set of n-tuples of the members of **I**.

If we know ϕ and if we know the usual rules holding of satisfaction (truth), we can in principle determine the truth-values of all the sentences of our first-order language. This is the cash value of the statement made above that the extensions of our individual constants and constant predicates are virtually all that we need in the theory of meaning in an applied first-order language.[6]

These conditions may be looked upon in slightly different ways. If ϕ is considered as an arbitrary function in (1.1)–(1.2), instead of that particular function which is involved in one's understanding of a language, and if **I** is likewise allowed to vary, we obtain a characterization of the concept of interpretation in the general model-theoretic sense.

III. PROPOSITIONAL ATTITUDES

We have to keep in mind the possibility that ϕ might be only a partial function (as applied to free singular terms), i.e., that some of our singular terms are in fact empty. This problem is not particularly prominent in the present paper, however.[7] If what I have said so far is correct, then the emphasis philosophers have put on the distinction between reference and meaning (e.g. between *Bedeutung* and *Sinn*) is motivated only in so far as they have implicitly or explicitly considered concepts which go beyond the expressive power of first-order languages.[8] Probably the most important type of such concept is a propositional attitude.[9] One purpose of this paper is to sketch some salient features of a semantical theory of such concepts. An interesting problem will be the question as to what extent we have to assume entities other than the usual individuals (the members of **I**) in order to give a satisfactory account of the meaning of propositional attitudes. As will be seen, what I take to be the true answer to this question is surprisingly subtle, and cannot be formulated by a simple 'yes' or 'no'.

What I take to be the distinctive feature of all use of propositional attitudes is the fact that in using them we are considering more than one possibility concerning the world.[10] (This consideration of different possibilities is precisely what makes propositional attitudes propositional, it seems to me.) It would be more natural to speak of different possibilities concerning our 'actual' world than to speak of several possible worlds. For the purpose of logical and semantical analysis, the second locution is much more appropriate than the first, however, although I admit that it sounds somewhat weird and perhaps also suggests that we are dealing with something much more unfamiliar and unrealistic than we are actually doing. In our sense, whoever has made preparations for more than one course of events has dealt with several 'possible courses of events' or 'possible worlds'. Of course, the possible courses of events he considered were from his point of view so many alternative courses that the actual events might take. However, only one such course of events (at most) became actual. Hence there is a sense in which the others were merely 'possible courses of events', and this is the sense on which we shall try to capitalize.

Let us assume for simplicity that we are dealing with only one propositional attitude and that we are considering a situation in which it is attributed to one person only. Once we can handle this case, a generalization to the others is fairly straightforward. Since the person in question remains constant throughout the first part of our discussion, we need not always indicate him explicitly.

IV. PROPOSITIONAL ATTITUDES AND 'POSSIBLE WORLDS'

My basic assumption (slightly oversimplified) is that an attribution of any propositional attitude to the person in question involves a division of all the possible worlds (more precisely, all the possible worlds which we can distinguish in the part of language we use in making the attribution) into two classes: into those possible worlds which are in accordance with the attitude in question and into those which are incompatible with it. The meaning of the division in the case of such attitudes as knowledge, belief, memory, perception, hope, wish, striving, desire, etc. is clear enough. For instance, if what we are speaking of are (say) *a*'s memories, then these possible worlds are all the possible worlds compatible with everything he remembers.

There are propositional attitudes for which this division is not possible. Some such attitudes can be defined in terms of attitudes for which the assumptions do hold, and thus in a sense can be 'reduced' to them. Others may fail to respond to this kind of attempted reduction to those 'normal' attitudes which we shall be discussing here. If there really are such recalcitrant propositional attitudes, I shall be glad to restrict the scope of my treatment so as to exclude them. Enough extremely important notions will still remain within the purview of my methods.

There is a sense in which in discussing a propositional attitude, attributed to a person, we can even restrict our attention to those possible worlds which are in accordance with this attitude.[11] This may be brought out e.g. by paraphrasing statements about propositional attitudes in terms of this restricted class of all possible worlds. The following examples will illustrate these approximate paraphrases:

> *a* believes that p = in all the possible worlds compatible with what *a* believes, it is the case that p;
>
> *a* does not believe that p (in the sense 'it is not the case that *a* believes that p') = in at least one possible world compatible with what *a* believes it is not the case that p.

V. SEMANTICS FOR PROPOSITIONAL ATTITUDES

What kind of semantics is appropriate for this mode of treating propositional attitudes? Clearly what is involved is a set Ω of possible worlds or of models in the usual sense of the word. Each of them, say $\mu \in \Omega$, is characterized by a set of individuals $\mathbf{I}(\mu)$ existing in that 'possible world'. An interpretation of individual constants and predicates will now be a two-argument function $\phi(a, \mu)$ or $\phi(Q, \mu)$ which depends also on the possible world μ in question. Otherwise an interpretation works in the same way as in the pure first-order case, and the same rules hold for propositional connectives as in this old case.

Simple though this extension of the earlier semantical theory is, it is in many ways illuminating. For instance, it is readily seen that in many cases earlier semantical rules are applicable without changes. *Inter alia*, in so far as no words for propositional attitudes occur inside the scope of a quantifier, this quantifier is subject to the same semantical rules (satisfaction conditions) as before.

VI. MEANING AND THE DEPENDENCE OF REFERENCE ON 'POSSIBLE WORLDS'

A new aspect of the situation is the fact that the reference $\phi(a, \mu)$ of a singular term now depends on μ – on what course the events will take, one might say. This enables us to appreciate an objection which you probably felt like making earlier when it was said that in a first-order language the theory of meaning is the theory of reference. What really determines the meaning of a singular term, you felt like saying, is not whatever reference it happens to have, but rather the way in which this reference is determined. But in order for this to make any difference, we must consider more than one possibility as to what the reference is, depending on the circumstances (i.e. depending on the course events will take). This dependence is just what is expressed by $\phi(a, \mu)$ when it is considered as a function of μ. (This function *is* the meaning of a, one is tempted to say.) Your objection thus has a point. However, it does not show that more is involved in the theory of meaning for first-order languages than the references of its terms. Rather, what is shown is that in order to spell out the idea that the meaning of a term is the way in which its reference is determined we have to consider how the reference varies in different possible worlds, and therefore go beyond first-order languages, just as I suggested above. Analogous remarks apply of course to the extensions of predicates.

Another novelty here is the need of picking out one distinguished possible world from among them all, viz. the world that happens to be actualized ('the actual world').

VII. DEVELOPING AN EXPLICIT SEMANTICAL THEORY: ALTERNATIVENESS RELATIONS

How are these informal observations to be incorporated into a more explicit semantical theory? According to what I have said, understanding attributions of the propositional attitude in question (let us assume that this is expressed by 'B') means being able to make a distinction between two kinds of possible worlds, according to whether they are compatible with the relevant attitudes of the person in question. The semantical counterpart to this is of course a function which to a given individual person assigns a set of possible worlds.

However, a minor complication is in order here. Of course, the person in question may himself have different attitudes in the different worlds we are considering. Hence this function in effect becomes a relation which to a given individual *and to a given possible world* μ associates a number of possible worlds which we shall call the *alternatives* to μ. The relation will be called the alternativeness relation. (For different propositional attitudes, we have to consider different alternativeness relations.) Our basic apparatus does not impose many restrictions on it. The obvious requirement that ensues from what has been said is the following:

(S.B.) $B_a p$ is true in a possible world μ if and only if p is true in all the alternatives to μ.

$B_a p$ may here be thought of as a shorthand for '*a* believes that *p*'. We can write this condition in terms of an interpretation function ϕ. What understanding B means is to have a function ϕ_B which to a given possible world μ and to a given individual a associates a set of possible worlds $\phi_B(a, \mu)$, namely, the set of all alternatives to μ.[12] Intuitively, they are the possible worlds compatible with the presence of the attitude expressed by B in the person a in the possible world μ.

In terms of this extended interpretation function, (S.B) can be written as follows:

$B_a p$ is true in μ if and only if p is true in every member of $\phi_B(a, \mu)$.

VIII. RELATION TO QUINE'S CRITERION OF COMMITMENT

The interesting and important feature of this truth-condition is that it involves quantification over a certain set of possible worlds. By Quine's famous criterion, we allegedly are ontologically committed to whatever we quantify over.[13] Thus my semantical theory of propositional attitudes seems to imply that we are committed to the existence of possible worlds as a part of our ontology.

This conclusion seems to me false, and I think that it in fact constitutes a counter-example to Quine's criterion of commitment *qua* a criterion of ontological commitment. Surely we must in some sense be *committed* to whatever we quantify over. To this extent Quine seems to be entirely right. But why call this a criterion of *ontological* commitment? One's ontology is what one assumes to exist in one's world, it seems to me. It is, as it were, one's census of one's universe. Now such a census is meaningful only in some particular possible world. Hence Quine's criterion can work as a criterion of *ontological* commitment only if the quantification it speaks of is a quantification over entities belonging to some one particular world. To be is perhaps to be a value of a bound variable. But to exist in an ontologically relevant sense, to be a part of the furniture of the world, is to be a value of a special kind of a bound variable, namely one whose values all belong to the same possible world. Thus the notion of a possible world serves to clarify considerably the idea of ontological commitment so as to limit the scope of Quine's dictum.

Clearly, our quantification over possible worlds does not satisfy this extra requirement. Hence there is a perfectly good sense in which we are not ontologically committed to possible worlds, however important their role in our semantical theory may be.

Quine's distinction between *ontology* and *ideology*, somewhat modified and put to a new use, is handy here.[14] We have to distinguish between what we are committed to in the sense that we believe it to exist in the actual world or in some other possible world, and what we are committed to as a part of our ways of dealing with the world conceptually, committed to as a part of our conceptual system. The former constitute our ontology, the latter our 'ideology'. What I am suggesting is that the possible worlds we have to quantify over are a part of our ideology but not of our ontology.

The general criterion of commitment is a generalization of this. Quantification over the members of one particular world is a measure of ontology, quantification that crosses possible worlds is often a measure of ideology. Quine's distinction thus ceases to mark a difference between two different types of studies or two different kinds of entities within one's universe. It now marks, rather, a distinction between the objects of reference and certain aspects of our own referential apparatus. Here we can perhaps see what the so-called distinction between theory of reference and theory of meaning really amounts to.

It follows, incidentally, that if we could restrict our attention to *one* possible world only, Quine's restriction would be true without qualifications. Of course, the restriction is one which Quine apparently would very much like to make; hence he has a legitimate reason for disregarding the qualifications for his own purposes.

Our 'ideological' commitment to possible worlds other than the actual one is neither surprising nor disconcerting. If what we are dealing with are the things people do – more specifically, the concepts they use – in order to be prepared for more than one eventuality, it is not at all remarkable that in order to describe these concepts fully we have to speak of courses of events other than the actual one.

IX. SINGULAR TERMS AND QUANTIFICATION IN THE CONTEXT OF PROPOSITIONAL ATTITUDES

Let us return to the role of individual constants (and other singular terms). Summing up what was said before, we can say that what the understanding of an individual constant amounts to in a first-order language is knowing which individual it stands for. Now it is seen that in the presence of propositional attitudes this statement has to be expanded to say that one has to know what the singular term stands for in the different possible worlds we are considering.

Furthermore, in the same way as these individuals (or perhaps rather the method of specifying them) may be said to be what is 'objectively given' to us when we understand the constant, in the same way what is involved in the understanding of a propositional attitude is precisely that distinction which in our semantical apparatus is expressed by the function which serves to define the alternativeness relation. This function is what is 'objectively given' to us with the understanding of a word for a propositional attitude.

These observations enable us to solve almost all the problems that relate to the use of identity in the context of propositional attitudes. For instance, we can at once see why the familiar principle of the substitutivity of identity is bound to fail in the presence of propositional attitudes when applied to arbitrary singular terms.[15] Two such terms, say a and b, may refer to one and the same individual in the actual world ($\phi(a, \mu_0)= \phi(b, \mu_0)$ for the world μ_0 that happens to be actualized), thus making the identity '$a=b$' true, and yet fail to refer to the same individual in some other (alternative) possible world. (I.e., we may have $\phi(a, \mu_1)\neq\phi(b, \mu_1)$ for some $\mu_1 \in \phi_B(c, \mu_0)$ where c is the individual whose attitudes are being discussed and B the relevant attitude.) Since the presence of propositional attitudes means (if I am right) that these other possible worlds have to be discussed as well, in their presence the truth of the identity '$a=b$' does not guarantee that the same things can be said of the references of a and b without qualification, i.e., does not guarantee the intersubstitutivity of the terms a and b.

Our observations also enable us to deal with quantification in contexts governed by words for propositional attitudes as long as we do not quantify *into* them. However, as soon as we try to do so, all the familiar difficulties which have been so carefully and persuasively presented by Quine and others will apply with full force.[16] An individual constant occurring within the scope of an operator like B which expresses a propositional attitude does not specify a unique individual. Rather, what it does is to specify an individual in each of the possible worlds we have to consider. Replace it by an individual variable, and you do not get anything that you could describe by speaking of the individuals over which this variable ranges. There are (it seems) simply no uniquely defined individuals here at all.

It is perhaps thought that the way out is simply to deny that one can ever quantify into a non-extensional context. However, this way out does not work.[17] As a matter of fact, in our ordinary language we often quantify into a grammatical construction governed by an expression for a propositional attitude. Locutions like 'knows who', 'sees what', 'has an opinion concerning the identity of' are cases in point, and so is almost any (other) construction in which pronouns are allowed to mix with words for propositional attitudes. Beliefs about 'oneself' and 'himself' yield further examples, and an account of their peculiarities leads to an interesting reconstruction of the traditional distinction between so-called modalities *de dicto* and *de re*.[18]

Another general fact is that we obviously have beliefs about definite individuals and not just about whoever happens to meet a certain description. I want to suggest that such beliefs (and the corresponding attitudes in the case of other propositional attitudes) are precisely what one half of the *de dicto – de re* distinction amounts to.[19]

Furthermore, it does not do to try to maintain that in these constructions the propositional attitude itself has to be taken in an unusual extensional or 'referentially transparent' sense. Such senses can in fact be defined in terms of the normal senses of propositional attitudes. However, these definitions already involve the objectionable quantification into opaque contexts, and if one tries to postulate the defined senses as irreducible primitive senses, they do not have the properties which they ought to have in order to provide the resulting quantified statements with the logical powers they in fact have in ordinary language. For instance, Quine's attempt to postulate a sense of (say) knowledge in which one is allowed to quantify into a context governed by a transparently construed construction 'knows that' has the paradoxical result that

$$(Ex) \text{ Jones knows that } (x=a)$$

is implied by any (transparently interpreted) statement of the form

Jones knows that $(b=a)$

and even by a similarly interpreted sentence

Jones knows that $(a=a)$.

This I take to show that the first of these three sentences can scarcely serve as a formulation of 'Jones knows who (or what) *a* is' in our canonical idiom. Yet no other paraphrase of this ubiquitous locution has been proposed, and none is likely to be forthcoming. (For what else can there be to Jones' knowing who *a* is than his knowing of some well-defined individual that Jones is that very individual? And it is Quine who always insists as strongly as anyone else that the values of bound variables have to be well-defined individuals.) It is not much more helpful to try to maintain that no true sentences of the form

Jones knows that $(b=a)$

(with the transparent sense of 'knows') are forthcoming whenever Jones fails to know who *a* is. The transparent sense in which this would be the case has never been explained in a satisfactory way, and I do not see how it can be done in a reasonable way without falling back to my own analysis. (What can it conceivably mean e.g. for Jones not to know in the transparent sense that an *a*, whom he knows to exist, is not self-identical? Can this self-identity fail to be true in a possible world compatible with everything Jones knows?)

Hence we have to countenance quantification into a context governed by an expression for an (opaquely construed) propositional attitude. Our semantical theory at once suggests a way of handling these problems. For instance, in order for existential generalization to be applicable to a singular term *b* occurring, say, in a context where *a*'s beliefs are being discussed, it has to be required that *b* refers to the *same* individual in the different possible worlds compatible with what *a* believes (plus, possibly, in the actual world). This, naturally, will be expressed by a statement of the form

(*) $(Ex)\,[B_a(x=b)\,\&\,(x=b)]$

or, if we do not have to consider the actual world, of the form $(Ex)\,B_a(x=b)$.

X. METHODS OF CROSS-IDENTIFICATION

This solution is simple, straightforward, and workable. It generalizes easily to other propositional attitudes. However, it hides certain interesting conceptual presuppositions. With what right do we speak of individuals in the different possible worlds as being *identical*? This is the problem to which we have to address ourselves.

It is not difficult to see what more there is given to us with our ordinary understanding of propositional attitudes that we have not yet dealt with. For instance, consider a man who has a number of beliefs as to what will happen tomorrow to himself and to his friends. Consider, on his behalf, a number of possible courses of events tomorrow. If I know what our man believes, I can sort these into those which are compatible with his beliefs as distinguished from those which are incompatible with them. But this is not all that is involved. Surely the same or largely the same individuals must figure in these different sequences of events. Under different courses of events a given individual may undergo different experiences, entertain different beliefs and hopes and fears; he may behave rather differently and perhaps even look somewhat different. Nevertheless our man can be (although he need not be) and usually is completely confident that, whatever may happen, he is going to be able to recognize (re-identify) his friends under these various courses of events, at least in principle. He may admit that courses of events are perhaps logically possible under which he would fail to do so; but these would not be compatible with his beliefs as to what will happen. Given full descriptions of two different courses of events tomorrow, both compatible with what our man believes ('believes possible', we sometimes say with more logical than grammatical justification), he will be able to recognize which individuals figuring in one of these descriptions are identical with which individual in the other, even if their names are being withheld. (Of course our man need not believe all this but my point is merely that he *can* and very often *does* believe it.)

The logical moral of this story is that together with the rest of our beliefs we are often given something more than we have so far incorporated into our semantical theory. We are given ways of *cross-identifying* individuals, that is to say, ways of understanding questions as to whether an individual figuring in one possible world is or is not identical with an individual figuring in another world.[20]

This is one point at which the obviousness of my claim may be partially obscured by my terminology. Let us recall what these 'possible worlds' are in the case of a propositional attitude. They are normally possible states of affairs or courses of events compatible with the attitude in question in some specified person. Now normally these attitudes may be attitudes toward definite persons or definite physical objects. But how is it that we may be sure, sight unseen, that the attitudes are directed toward the right persons or objects? Only if in all the possible worlds compatible with the attitude in question we can pick out the recipient of this attitude, i.e. the individual at its receiving end. Although in many concrete situations the possibility of doing so is obvious, it has not been built into our semantical apparatus so far. There is so far nothing in our semantical theory which enables us to relate to each other the members of the different domains of individuals $\mathbf{I}(\mu)$. In many, though not necessarily all, applications of such relations are given to us as a part of our understanding of the concepts involved. For such cases, we have to build a richer semantical theory.

The way to do so is to postulate a method of making cross-identifications. One possible way to do so is to postulate a set of functions **F** each member f of which picks out at most one individual $f(\mu)$ from the domain of individuals $\mathbf{I}(\mu)$ of each given model μ. We must allow that there is no such value for some models μ. In other words, $f \in \mathbf{F}$ may be a partial function. Furthermore, we must often require that, given $f_1, f_2 \in \mathbf{F}$, if $f_1(\mu) = f_2(\mu)$ then $f_1(\lambda) = f_2(\lambda)$ for all alternatives λ to μ. In other words, an individual cannot 'split' when we move from a world to its alternatives. This question may seem to be a mere matter of detail, but it is easily seen that the question whether an individual can split in the sense just explained is tantamount to the question whether the substitutivity of identity can fail for bound (individual) variables, i.e. to the question whether a sentence

$$(Ux)(Uy)(x = y \supset B_a(x = y))$$

can fail to be logically true. This, again, is tantamount to the question whether a sentence of the form

$$(Ux)(Uy)(x = y \supset (Q(x) \supset Q(y)))$$

(with just one layer of operators for propositional attitudes in Q) can fail to be logically true.

In terms of the set **F**, the question whether $a \in \mathbf{I}(\mu)$ is identical with $b \in \mathbf{I}(\lambda)$ amounts to the question whether there is a function of $f \in \mathbf{F}$ such that $f(\mu) = a, f(\lambda) = b$.

XI. THE ROLE OF INDIVIDUATING FUNCTIONS

Instead of speaking of a set of functions correlating to each other the individuals existing in the different possible worlds, it is often more appropriate to speak of these domains of individuals as being partly identical (overlapping). Then there would be no need to speak of correlations at all. This point of view is useful in that it illustrates the fact that the apparently different individuals which are correlated by one of the functions $f \in \mathbf{F}$ is just what we ordinarily mean by one and the same individual. It is the concrete individual which we speak about, which we give a name to, etc. In fact, the members of **F** might in fact be thought of as names or individual constants of a certain special kind, namely those having a unique reference in all the different worlds we are speaking of and hence satisfying formulas of the form (*). Indeed, I shall assume in the sequel that a constant of this kind can be associated with each function $f \in \mathbf{F}$.

However, emphasizing the role of the functions $f \in \mathbf{F}$ is useful for several purposes. First and foremost, it highlights an extremely important non-trivial part of our native conceptual skills, namely, our capacity to recognize one and the same individual under different circumstances and under different courses of events. What the set **F** of functions embodies is just the totality of ways of doing this. The non-trivial character of the possibility of this recognition would be lost if we should simply speak of the members of the different possible worlds as being partly identical.

For another thing, the structure formed by the relations of cross-world identity (David Kaplan calls them "trans world heir lines") may be so complex as to be indescribable by speaking simply of partial identities between the domains of individuals of the different possible worlds. Above, it was said that in the case of many propositional attitudes an individual cannot 'split' when we move from a world to its alternatives. Although this seems to me to be the case with all the propositional attitudes I have studied in any detail, it is not quite clear to me precisely why this should always be the case. At any rate, there seem to be reasons for suspecting that the opposite 'irregularity' can occasionally take place with some modalities: individuals can 'merge together' when we move from a world to its alternatives. An analogy with temporal modalities may be instructive here.[21] If we presuppose some suitable system of cross-identifications between individuals existing at different times which turn on continuity, it seems possible in principle that a singular term should refer to the same physical system at all the different moments of time we are considering although this system 'merges' with others at times and occasionally 'splits up' into several. Some of these complications seem to be impossible to rule out completely in the case of some propositional attitudes, and because of them the idea of partly overlapping domains seems to me seriously oversimplified.

An extremely important further reason why we cannot reify the members of **F** into ordinary individuals is the possibility of having two different methods of cross-identification between the members of the same possible worlds, i.e. two different sets of 'individuating functions' although we are dealing with precisely the same sets of possible worlds. I have argued elsewhere that this kind of situation is not only possible to envisage but is actually present in our own ways with perceptual contexts.[22] It would take us too far to show precisely what is involved in such cases. Suffice it to point out that this claim, if true, would strikingly demonstrate the dependence of our methods of cross-identification on our own conceptual schemes and hence on things of our own creation. The apparent simplicity of our idea of an 'ordinary' individual, safe as it may seem in its solid commonplace reality, is thus seen to be merely a reflection of the familiarity and relatively deep customary entrenchment of one particular method of cross-identification, which *sub specie aeternitatis* (i.e. *sub specie logicae*) nevertheless enjoys but a relative privilege as against a host of others.

The methods of cross-identification represented by the set **F** of 'individuating functions', as we might call them, also call for several further comments.

The main function of this part of our semantical apparatus is to make sense of quantification into contexts of propositional attitudes. The truth-conditions of statements in which this happens can be spelled out in terms of membership in **F**. As an approximation we can say the following: A sentence of the form $(Ex)\,Q(x)$ is true in μ if and only if there is an individual constant (say b) associated with some $f \in \mathbf{F}$ such that $Q(b)$ is true in μ. This approximation shows, incidentally, how close we can stick to the simple-minded idea that an existentially quantified sentence is true if and only if it has a true substitution instance. The only additional requirement we need is that the substitution-value of the bound variable has to be of the right sort, to wit, has to specify the same individual in all the possible worlds we are speaking of in the existential sentence in question. This is what is meant by the requirement that b has to be associated with one of the functions $f \in \mathbf{F}$.

This approximation, although not unrepresentative of the general situation, requires certain modifications in order to work in all cases. The set **F** has to be relativized somewhat in the same way the unrestricted notion of a possible world was relativized by the notion of alternative in the truth-criterion (S.B.) above. (Not everyone is in all situations 'familiar with' all the relevant methods of individuation, it might be said.) I shall not discuss the ensuing complications here, however, for they do not change the overall picture in those respects which are relevant in the rest of this paper.

XII. STATEMENTS ABOUT DEFINITE INDIVIDUALS VS. STATEMENTS ABOUT WHOEVER OR WHATEVER IS REFERRED TO BY A TERM

The possibility of quantifying across an operator which expresses a propositional attitude enables us to explicate the logic of the locutions in which we need this possibility in the first place. Perhaps the most important thing we can do here is to make a distinction between propositional attitudes directed to whoever (whatever) happens to be referred to by a term and attitudes directed toward a certain individual, independently of how he happens to be referred to. This distinction was hinted at above. Now it is time to explain it more fully. For instance, someone may have a belief concerning the next Governor of California, whoever he is or may be, say, that he will be a Democrat. This is different from believing something about the individual who, so far unbeknownst to all of us, in fact is the next Governor of California.

In formal terms, the distinction is illustrated by the pair of statements

$$B_a\,(g \text{ is a Democrat})$$
$$(Ex)\,((x = g)\,\&\,B_a\,(x \text{ is a Democrat})).$$

Notice, incidentally, that my way of drawing this distinction implies that one can have (say) a belief concerning the individual who in fact is *a* only if such an individual really exists, whereas one can in principle have a belief concerning *a*, 'whoever he is', even though there is no such person. This, of course, is just as it ought to be.

The naturalness of our semantical conditions, and their close relation to the realities of actual usage, can be illustrated by applying them to what I have called a statement about a definite individual. As an example, we can use

> '*a* believes of the man who in fact is Mr. Smith that he is a thief',

in brief,

$$(Ex)\,(x = \text{Smith}\,\&\,B_a\,(x \text{ is a thief})).$$

In order for this to be true, there has to be some $f \in \mathbf{F}$ such that the value of f in the actual world (call it μ_0) exists and is Smith and that $f(\mu)$ has the property of being a thief whenever $\mu \in \phi_B(a, \mu_0)$, i.e. in all the alternatives to the actual world.

What the requirement of the existence of f amounts to is clear enough. If it is true to say that *a* has a belief about *the particular individual* who in fact is Smith, then *a* clearly must believe that he can characterize this individual uniquely. In other words, he must have some way of referring to or characterizing this individual in such a way that one and the same individual is in fact so characterized in all the worlds compatible with what he believes. This is precisely what the existence of f amounts to. If no such function existed, *a* would not be able to pick out the individual who in fact is Smith under all the courses of events he believes possible, and there would not be any sense in saying that *a*'s belief is *about* the particular individual in question.

XIII. INDIVIDUATING FUNCTIONS VS. INDIVIDUAL CONCEPTS

One important consequence of my approach is that not every function which from each μ picks out an individual can be said to specify a unique individual. In fact, many perfectly good free singular terms fail to do so in the context of many propositional attitudes. Even proper names fail to do so in epistemic contexts, for one may fail to know who the bearer of a given proper name is.

Such arbitrary functions may be important for many purposes. They are excellent approximations in our theory to the 'individual concepts' which many philosophers have postulated.[23] (In Section VI above we already met a number of such 'individual concepts' in the form of the functions $\phi(a, \mu)$ with a fixed a.) Each such individual concept specifies or 'contains', as Frege would say, not just a reference (in the actual world) but also the way in which this reference is given to us. Each of them would thus qualify for a sense (*Sinn*) of a singular term *à la* Frege.[24] However, we do not need the totality of such arbitrary functions in the semantics which I am building up and which (I want to argue) is largely implicit in our native conceptual apparatus. Quine's criterion, however misleading it may be as a criterion of ontological commitment, still works as a criterion of commitment. If it is applied here, it shows that we are not committed (ontologically or 'ideologically') to these arbitrary functions, since we do not have to quantify over them, only over the members of the much narrower class **F**.

The other side of the coin is that in our semantical apparatus we do have to quantify over the members of **F**. Does it follow that they 'exist' or 'are part of our ontology'? An answer to this question can be given along the same lines as to the corresponding question concerning 'possible worlds'. The members of **F** are not members of any possible world; they are not part of anybody's count of 'what there is'. They may 'subsist' or perhaps 'exist', and they are certainly 'objective', but they do not have any ontological role to play. The need to distinguish between ontology and 'ideology' is especially patent here.

The functions that belong to **F** may of course be considered special cases of the 'individual concepts' postulated by some philosophers of logic or as special cases of Frege's 'senses' (*Sinne*). No identification is possible between the two classes, however, for we saw earlier that not every arbitrary singular term (say b) which picks out an individual from each $\mathbf{I}(\mu)$ we are considering goes together with an $f \in \mathbf{F}$, although every such term is certainly meaningful and hence has a Fregean 'sense' and perhaps even gives us an 'individual concept'. As I have put it elsewhere, members of **F** do not only involve a 'way of being given' as Frege's senses do, but also *a way of being individuated*.[25] The primary care is in our approach devoted to ordinary concrete individuals. Singular terms merit a special honorary mention only if they succeed in picking out a unique individual of this sort.

Let us say that an $f \in \mathbf{F}$ is (gives us) an individuating concept, and let us say that a term b does *individuate* (in the context of discussing a's beliefs) in so far as

$$(Ex)\ B_a\ (x = b)$$

is true. Then we could have individuation without reference and reference without individuation: Both

$$(Ex)\ B_a\ (x = b)\ \&\sim (Ex)\ (x = b)$$

and

$$(Ex)\ (x = b)\ \&\sim (Ex)\ B_a\ (x = b)$$

can be true. We could even have both, but without matching:

$$(Ex)\ (x = b)\ \&(Ex)\ B_a\ (x = b)\ \&\sim (Ex)\ ((x = b)\ \&\ B_a\ (x = b))$$

is satisfiable. Only if

$$(Ex)\ ((x = b)\ \&\ B_a\ (x = b))$$

is true does the successful individuation give us the individual which the term b actually refers to.

XIV. THE THEORY OF REFERENCE AS REPLACING THE THEORY OF MEANING

Here we are perhaps beginning to see what I meant when I said at the beginning of this paper that what is often called the theory of meaning is better thought of as the theory of reference for certain more complicated conceptual situations. Some of the most typical concepts used in the theory of meaning, such as Frege's *Sinne* and the 'individual concepts' of certain other philosophers of logic were in the first place introduced to account for such puzzles as the failure of the substitutivity of identity and the difficulty of quantifying into opaque contexts (e.g. into a context governed by a word for a propositional attitude). I have argued, however, that a satisfactory semantical theory which clears up these puzzles can be built up without using Frege's *Sinne* and without any commitment to individual concepts in any ordinary sense of the word. Instead, what we need are the individuating functions, i.e., the members of **F**. And what these functions do is not connected with the ideas of the traditional theory of meaning. What they do is precisely to give us the individuals which we naively think our singular statements to be about and which we think our singular terms as referring to. This naive point of view is essentially correct, it seems to me. The functions of $f \in \mathbf{F}$ are the prime vehicles of our references to individuals when we discuss propositional attitudes. What is not always realized, however, is how much goes into our ordinary concepts of an individual and a reference. These are not specified in a way which works only under one particular course of events. They are in fact specified in a way which works under a wide variety of possible courses of events. But, in order to spell out this idea, we are led to consider several possible worlds, with all the problems with which we have dealt in this paper, including very prominently the problem of cross-identifying individuals.

The function of our 'individuating functions', i.e. the members of the set **F**, is to bring out these hidden – or perhaps merely overlooked – aspects of our concept of an individual (definite individual). This close connection between the set **F** and the concept of an individual appears in a variety of ways. One may for instance think of the role which the membership in **F** plays in the truth-conditions which we set up above for quantification into modal contexts. When it is asked in such a context whether there exists an individual of a certain kind, a singular term specifies such an individual only if its references match the values of a unique member of **F** in all the relevant possible worlds. Thus it is these functions that in effect give us the individuals which can serve as values of bound variables. As we saw above, it is mainly the possible subtlety and multiplicity of relations of cross-identity that prevent us from simply making the domains of the different possible worlds partly identical and thus hypostatizing my individuating functions into commonplace individuals.

This connection between individuating functions and the concept of an individual is part of what justifies us in thinking that in the traditional dichotomy their theory would belong primarily to the theory of reference, in spite of the fact that their main function in our semantical theory is to solve some of the very problems which the traditional theory of meaning was calculated to handle. This role is perhaps especially clear in connection with the substitutivity of identity. As we have seen, this principle does not hold for arbitrary singular terms *a*, *b*. However, if it is required in addition that both of these terms specify a well-defined individual, i.e. satisfy expressions like (*), depending on the context, then the substitutivity of identicals is easily seen to hold, presupposing of course here the prohibition against splitting that was mentioned above. What this observation shows is clear enough. The failure of the substitutivity of identity poses one of the most typical problems for the treatment of which meanings, individual concepts and other paraphernalia of the theory of meaning were introduced in the first place. If the substitutivity of identity fails, clearly we cannot be dealing with ordinary commonplace individuals, it was alleged, for if two such individuals are in fact identical, surely precisely same things can be said of them. This is what prompts the quest for individuals of some non-ordinary sort, capable of restoring the substitutivity principle when used as references of our terms. (This is almost precisely Frege's strategy.) We have seen, however, that the (apparent) failure of the substitutivity is due simply to the failure of some free singular terms to specify the same individual in the different 'possible worlds' we have to consider. Moreover, we have seen that this apparent failure is automatically corrected in precisely those cases in which it ought to be corrected, viz. in the cases where the two terms in question really do specify a *unique* individual. (That this depends on certain specific requirements concerning our methods of cross-identification, viz. on a prohibition against 'splitting', does not affect my point.) Substitutivity of identity is restored, in brief, not by requiring that our singular terms refer to the entities postulated by the so-called theory of meaning, but by requiring (in the form of an explicit premise) that they really succeed in specifying uniquely the kind of ordinary individual with which the theory of reference typically deals. One can scarcely hope to find a more striking example of the breakdown of the distinction between a theory of meaning and a theory of reference.[26]

XV. TOWARDS A SEMANTIC NEOKANTIANISM

The aspect of my observations most likely to upset many contemporary philosophers is the ensuing implicit dependence of our concept of an individual on our ways of cross-identifying members of different 'possible worlds'. These 'possible worlds' and the supply of individuating functions which serve to interrelate their respective members may enjoy, and in my view do enjoy, some sort of objective reality. However, their existence is not a 'natural' thing. They may be as solidly objective as houses or books, but they are as certainly as these created by men (however unwittingly) for the purpose of facilitating their transactions with the reality they have to face. Hence my reasoning ends on a distinctly Kantian note. Whatever we say of the world is permeated throughout with concepts of our own making. Even such *prima facie* transparently simple notions as that of an individual turn out to depend on conceptual assumptions dealing with different possible states of affairs. As far as our thinking is concerned, reality cannot be in principle wholly disentangled from our concepts. A *Ding an sich*, which could be described or even as much as individuated without relying on some particular conceptual framework, is bound to remain an illusion.

REFERENCES

[1] See e.g. W. V. Quine, *From a Logical Point of View*, Harvard University Press, Cambridge, Mass., 1953 (2nd ed.: 1961), pp. 130–132.

[2] For a simple recent argument of this sort (without a specific reference to first-order theories), see e.g. William P. Alston, *Philosophy of Language*, Prentice-Hall, Inc., Englewood Cliffs, N.J., 1964, p. 13. Cf. also Quine, *op. cit.* pp. 21–22.

[3] In more general terms, it seems to me hopeless to try to develop a theory of sentential meaning which is not connected very closely with the idea of the information which the sentence can convey to us, or a theory of meaning for individual words which would not show how understanding them contributes to appreciating the information of the sentences in which they occur. There are of course many nuances in the actual use of words and sentences which are not directly explained by connecting meaning and information in this way, assuming that this can be done. However, there do not seem to be any obstacles in principle to explaining these nuances in terms of pragmatic, contextual, and other contingent pressures operating on a language-user. For remarks on this methodological situation, see my paper 'Epistemic Logic and the Methods of Philosophical Analysis', *Australasian Journal of Philosophy* **46** (1968) 37–51.

[4] Donald Davidson, 'Truth and Meaning', *Synthese* **17** (1967) 304–323.

[5] See Quine, *op. cit.*, pp. 32–37.

[6] The main reason why the truth of these observations is not appreciated more widely seems to be the failure to consider realistically what the actual use of a first-order language (say for the purpose of conveying information to another person) would look like.

[7] The basic problems as to what happens when this possibility is taken seriously are discussed in my paper, 'Studies in the Logic of Existence and Necessity I', *The Monist* **50** (1966) 55–76, reprinted (with changes) in the present volume under the title 'Existential Presuppositions and Their Elimination' (above pp. 23–44).

[8] This is certainly true of Frege. His very interest in oblique contexts seems to have been kindled by the realization that they cannot be handled by means of the ideas he had successfully applied to first-order logic.

[9] The term seems to go back to Bertrand Russell, *An Inquiry into Meaning and Truth*, George Allen and Unwin, London, 1940.

[10] An important qualification here is that for deep logical reasons one cannot usually distinguish effectively between what is 'really' a logically possible world and what

merely 'appears' on the face of one's language (or thinking) to be a possibility. This, in a sufficiently sharp analysis, is what destroys the pleasant invariance of propositional attitudes with respect to logical equivalence. Even though p and q are equivalent, i.e. even though the 'real' possibilities concerning the world that they admit and exclude are the same,

	knows	
	believes	
a	remembers	that p
	hopes	
	strives	

and

	knows	
	believes	
a	remembers	that q
	hopes	
	strives	

need not be equivalent, for the apparent (to a) possibilities admitted by p and q need not be identical.

I have studied this concept of an 'apparent' possibility and its consequences at some length elsewhere (especially in the second and third paper printed in *Deskription, Analytizität und Existenz* (ed. by Paul Weingartner), Pustet, Salzburg and Munich 1966, in 'Are Logical Truths Analytic?', *Philosophical Review* **74** (1965) 178–203, in 'Surface Information and Depth Information' forthcoming in *Information and Inference* (ed. by K. J. J. Hintikka and P. Suppes), D. Reidel Publishing Co., Dordrecht, 1969), and in 'Are Mathematical Truths Synthetic A Priori?', *Journal of Philosophy* **65** (1968) 640–651.

It is an extremely interesting concept to study and to codify. However, it is not directly relevant to the concerns of the present paper, and would in any case break its confines. Hence it will not be taken up here, except by way of this *caveat*.

[11] There is a distinction here which is not particularly relevant to my concerns in the present paper but important enough to be noted in passing, especially as I have not made it clear in my earlier work. What precisely are the worlds 'alternative to' a given one, say μ? A moment's reflection on the principles underlying my discussion will show, I trust, that they must be taken to be worlds compatible with a certain person's having a definite propositional attitude in μ, and not just compatible with the content of his attitude, for instance, compatible with someone's knowing something in μ and not just compatible with what he knows. I failed to spell this out in my *Knowledge and Belief* (Cornell Univ. Press, Ithaca, N.Y., 1962), as R. Chisholm in effect pointed out in his review article, 'The Logic of Knowing', *Journal of Philosophy* **60** (1963) 773–795.

[12] As the reader will notice, I am misusing (in the interest of simplicity) my terminology systematically by speaking elliptically of 'the person a' etc. when 'the person referred to by a' or some such thing is meant. I do not foresee any danger of confusion resulting from this, however.

[13] E.g. Quine, *op. cit.*, pp. 1–14, W.V. Quine, *Word and Object*, The MIT Press, Cambridge, Mass., and John Wiley, New York and London, 1960, pp. 241–243. It is not quite clear from Quine's exposition, however, precisely how much emphasis is to be put on the word 'ontology' in his criterion of ontological commitment. My discussion which focuses on this word may thus have to be taken as a qualification to Quine's criterion rather than as outright criticism.

[14] Quine, *From a Logical Point of View*, pp. 130–132.
[15] For a discussion of the problems connected with the substitutivity principle, see my exchange with Føllesdal: Dagfinn Føllesdal, 'Knowledge, Identity, and Existence', *Theoria* **33** (1967) 1–27; Jaakko Hintikka, 'Existence and Identity in Epistemic Contexts', *ibid.* 138–147.
[16] See Quine, *From a Logical Point of View*, ch. 8; *Word and Object*, ch. 6; *The Ways of Paradox and Other Essays*, Random House, New York, 1966, chapters 13–15.
[17] Some arguments to this effect were given in *Knowledge and Belief* (ref. 11 above), pp. 142–146. The only informed criticism of this criticism that I have seen has been presented by R. L. Sleigh, in a paper entitled 'A Note on an Argument of Hintikka's', *Philosophical Studies* **18** (1967) 12–14. As I point out in my reply, 'Partially Transparent Senses of Knowing' (forthcoming), Sleigh's argument turns on an ambiguity in my original formulation which is easily repaired. Neither the ambiguity nor its elimination provides any solace to the adherents of the view I have criticized, however.
[18] One thing at which this old distinction aims is obviously the distinction (which I am about to explain) between statements about whoever or whatever meets a description and statements about the individual who in fact does so. For the distinction, cf. Jaakko Hintikka, 'Individuals, Possible Worlds, and Epistemic Logic', *Nous* **1** (1967) 32–62, especially 46–49, as well as '"Knowing Oneself" and Other Problems in Epistemic Logic', *Theoria* **32** (1966) 1–13.
[19] Cf. below (Section XII).
[20] Cf. here my paper, 'On the Logic of Perception' in *Perception and Personal Identity* (ed. by N. Care and R. Grimm), Case Western Reserve Univ. Press, Cleveland, 1969. (Reprinted in the present volume, pp. 151–183.)
[21] For temporal modalities, see e.g. A. N. Prior, *Past, Present and Future*, Clarendon Press, Oxford, 1967. – I am not saying that our actual methods of cross-identification in the case of temporal modalities (i.e. on ordinary methods of re-identification) turn on continuity quite as exclusively as I am about to suggest. It suffices for my purposes to present an example of methods of cross-identification that allows both 'branching' and 'merging', and it seems to me at least conceivable that temporal modalities might under suitable circumstances create such a situation.
[22] This is argued in 'On the Logic of Perception' (ref. 20 above).
[23] Cf. e.g. R. Carnap, *Meaning and Necessity*, University of Chicago Press, Chicago, 1947 (2nd ed.: 1956), pp. 41, 180–181, and Section VI *supra.*
[24] Cf. Gottlob Frege, 'Über Sinn und Bedeutung', *Zeitschrift für Philosophie und philosophische Kritik* **100** (1892) 25–50, especially p. 26, last few lines.
[25] Cf. 'On the Logic of Perception' (ref. 20 above).
[26] Views closely resembling some of those which I am putting forward here (and in some cases anticipating them) have been expressed by David Kaplan, Richard Montague, Dagfinn Føllesdal, Stig Kanger, Saul Kripke, and others. Here I am not trying to relate my own ideas to theirs. It is only fair, however, to emphasize my direct and indirect debts to these writers.

EXISTENTIAL PRESUPPOSITIONS AND UNIQUENESS PRESUPPOSITIONS

I. EXISTENTIAL PRESUPPOSITIONS

The meeting at which this paper was first read was primarily devoted to presuppositionless logics, somewhat misleadingly known as free logics.[1] (The term misleads because of the absence of any connection between these 'free' logics and the well-known free algebras.) The presuppositions which these logics dispense with are presuppositions of existence, typically presuppositions to the effect that certain free singular terms are not empty.

I have some doubts as to how interesting an enterprise the study of such presuppositionless logics will turn out to be in the long run. It seems to me that a study of the role of existential presuppositions in first-order logic will fairly soon exhaust all the general theoretical interest that there is in the area, if conducted as a purely syntactical or semantical enterprise. It is true that there is a considerable variety of ways in which a semantics can be built for a presuppositionless first-order logic. However, a philosophically satisfactory comparison between them will in my opinion have to turn on a deeper conceptual analysis of the situation than the standard syntactical and semantical methods afford.

There are at least two main directions into which the study of presupposition-free logics can be developed so as to preserve its continued interest. One main direction which needs – and deserves – much more discussion than it has so far received is the deeper philosophical analysis of the concept of existence, on which the concept of existential presupposition is based. My original intention was to discuss a line of approach to this area which looks especially interesting. It seems to me that we have to go beyond the usual syntactical and semantical tools in order to get at the interesting problems here. What we study in semantics are roughly speaking a language plus the representative relations that connect it with its several interpretations.[2] What we do not study in semantics is what goes into these relations, that is, the dynamics of the interplay of language and reality when an applied language is actually being used to some purpose. Elsewhere,[3] I have suggested what the 'language-games' look like which go together with quantifiers – and therefore also with the concept of existence which is codified by the existential quantifier. I have argued that in this case the use of the frequently misapplied label 'language-game' need not give rise to any compunctions, for the games in question can be so called in the precise sense of the mathematical theory of games. I have even ventured to suggest that this game-theoretical interpretation of first-order logic can serve as a starting-point for interesting philosophical and logical theorizing. In order to be able to continue the line of thought which the preceding papers represent, I have nevertheless decided not to take up here these in my judgment highly important matters, and decided to try to say instead something that seems to be more closely related to the other papers of this volume.

II. MODAL LOGICS AS 'FREE' LOGICS

A second main enterprise which seems very promising is to examine presuppositions of existence in languages richer than those studied in first-order logic. Among such richer languages, those using modal notions (including propositional attitudes) seem to be particularly interesting.

It seems to me that this is a most promising area for further work, although it also seems to me that most of the crucial insights required to see our way through have already been obtained. However, many of the consequences of these insights still remain in the dark. I surmise that future work in this area will have extremely important general philosophical implications. The problems to which this work seems relevant include such questions as e.g. the question as to what entities are needed in a satisfactory semantics for modal logics, the distinction between a theory of meaning and a theory of reference, and an evaluation of Quine's criticism of quantified modal logic.

Not all work in this direction may at first seem related to existential presuppositions. In this paper, I shall nevertheless argue that the presuppositions of uniqueness whose failure is responsible for the puzzling breakdown of some of the most characteristic laws of first-order logic (notably of the law of existential generalization) in modal contexts are to a surprising extent analogous to the presuppositions of existence which we need in first-order logic. Even the explicit formulation of these presuppositions in one's system of modal logic can be elicited in the same way as the famous 'Quinean' formulation '$(Ex)(x=b)$' of the existential presuppositions of first-order logic.[4] Hence almost any serious study of the semantics of modal logic will be related very closely to the study of logics without existential presuppositions. What one needs in the theory of modal logics is in my opinion quite literally a presuppositionless logic (or 'free logic', to use this unfortunate label), namely, a logic without presuppositions of uniqueness. To coin a slogan, modal logics are really only so many 'free' (i.e. presuppositionless) logics.

Since this point of fundamental importance to our general view of the whole field of presuppositionless logics and of its prospects, I shall try to argue for it, even on pain of occasionally repeating what I have said elsewhere.[5]

III. MODEL SETS AND MODEL SYSTEMS

I shall first discuss a quantified modal logic with '(Ex)', '(Ux)' as quantifiers and 'M', 'N' as the two (dual) modal operators. (I assume that they correspond, roughly, to necessity and possibility, but I shall leave their precise interpretations unspecified.) As propositional connectives, I shall employ '$\sim$', '&', '$\vee$', and I shall normally assume (for simplicity) that all negation-signs have been driven into the formulas so as to be prefixed to atomic ones or to identities.

I shall use the familiar framework of model sets and model systems. Dagfinn Føllesdal has criticized me repeatedly for speaking of them and not of the 'real' models and systems of models.[6] However, while I may have given the impression of overestimating the ease at which considerations pertaining to model sets can be transposed so as to apply to models, I think that he is still underestimating it. Later in the present paper, I shall say a few things about the relation of my methods to the usual semantical methods which hopefully illustrate this ease.

The conditions that define a model set (say μ) in first-order logic are well known, but still brief enough to be reproduced here:

(C.$\sim$) Not $p\in\mu$, $\sim p\in\mu$ (p atomic or an identity).
(C.&) If $(p \,\&\, q)\in\mu$, then $p\in\mu$, $q\in\mu$.
(C.$\vee$) If $(p \vee q)\in\mu$, then either $p\in\mu$ or $q\in\mu$ (or both).

(C.E) If $(Ex)p \in \mu$, then $p(a/x) \in \mu$ for at least one singular constant 'a'.

(C.U) If $(Ux)p \in \mu$, and if 'b' occurs in the formulas of μ, then $p(b/x) \in \mu$.

(C.self≠) Not '$(b \neq b)$' $\in \mu$.

(C.=) If p is atomic or an identity, if $p(a/b) = q(a/b)$, $p \in \mu$ and '$(a = b)$' $\in \mu$, then $q \in \mu$.

In (C.E), $p(a/x)$ is the result of substituting 'a' for 'x' everywhere in p. The same notation is used in (C.U) and (C.=), and will be used in what follows. In (C.self≠), '$(b \neq b)$' is of course a shorthand for '$\sim (b = b)$'.

A model system Ω is a set of model sets with a two-place relation (called the alternativeness relation) defined on it in such a way that the following conditions are satisfied:

(C.M*) If $Mp \in \mu \in \Omega$, there is an alternative $\lambda \in \Omega$ to μ such that $p \in \lambda$.

(C.N$^+$) If $Np \in \mu \in \Omega$ and if $\lambda \in \Omega$ is an alternative to μ, then $p \in \lambda$.

(C.refl) The alternativeness relation is reflexive.

Instead of (C.refl), we may want to adopt for some applications the weaker condition:

(C.n*) If $Np \in \mu \in \Omega$, there is at least one alternative $\lambda \in \Omega$ to μ such that $p \in \lambda$.

The satisfiability of a set of formulas λ is defined as its imbeddability in some member $\mu \supseteq \lambda$ of a model system Ω, $\mu \in \Omega$. When in the sequel I shall consider changes in the conditions defining a model system, I consider them substantial only when they affect the concept of satisfiability so defined.

A formula is said to be logically true if the unit set of its negation is not satisfiable.

Independently of what one thinks of the relation of my conditions to semantics proper, a certain plausibility cannot be denied of them, it seems to me. Model sets can obviously be thought of as (partial) descriptions of possible worlds, and alternatives to a member (say μ) of a model system can be thought of as descriptions of worlds 'alternative to' the one described by μ (in the case of necessity and possibility, of worlds that in some relevant sense could have been realized instead of the one described by μ). The basic intuitive idea is, roughly, that necessity (in a given possible world) equals truth in all alternative possible worlds, and that possibility therefore equals truth in at least one alternative possible world.

IV. THE BEHAVIOR OF IDENTITY IN MODAL CONTEXTS

(C.$\sim$) can be strengthened so as to omit the restriction to atomic formulas and identities. However, (C.$=$) cannot be so modified without affecting the concept of satisfiability and without creating paradoxical applications (when arbitrary free singular terms are substituted for our individual constants).

However, it can be shown that a weaker principle of substitutivity is a consequence of our conditions:

(C.N$=$) If $p \in \mu$, if p results from q by interchanging 'a' and 'b' in a number of places which are within the scope of precisely $n_2, n_2, \ldots$ modal operators, respectively, and if '$N^{n_1}(a=b)$'$\in \mu$, '$N^{n_2}(a=b)$'$\in \mu, \ldots$, then $q \in \mu$.

Here 'N^{n_1}' is of course a shorthand for '$NN \ldots N$' (n_1 occurrences of N), and similarly for the other n_i.

It can be shown that this holds independently of (C.refl). If this condition is assumed, then instead of the several iterated necessary identities we need only '$N^{n_0}(a=b)$'$\in \mu$ where $n_0 = \max(n_1, n_2, \ldots)$. The case $n_0 = 0$ is included in our formulation as a special case.

The semantic import of (C.N$=$) is clear. When a singular term occurs outside the scope of modal operators, we are speaking merely of its reference in the actual world. When one of its occurrences is buried under n_i layers of modal operators, we are speaking of its several references in the possible worlds described by all the different alternatives, n_i times removed, to the description of the actual one. In order for the terms 'a' and 'b' to be interchangeable at such an occurrence, they must refer to the same individual in each of these possible worlds. That this is the case is precisely what '$N^{n_i}(a=b)$' expresses. Thus (C.N$=$) is very plausible even without the possibility of reducing it to the simpler condition (C.$=$). No wonder, therefore, that occasionally it has been said that principles like (C.N$=$) are the form which the principle of the substitutivity of identity takes in modal contexts. It can be shown, however, that (C.N$=$)

follows from the other conditions (in the sense that it does not affect the notion of satisfiability), and hence does not embody any assumptions not already made in ordinary non-modal first-order logic.

A proof that the replacement of (C.=) by (C.N=) does not affect satisfiability is straightforward.[7]

If (C.=) is not restricted to atomic formulas and identities, various paradoxes will ensue. I shall not discuss them here, however, for the restriction already gives us a satisfactory (and intuitive) way of avoiding all the difficulties. A brief discussion of this subject is given in the last couple of pages of 'Modality and Quantification' in the present volume, pp. 57–70.

It might be thought that my giving up of the stronger (unrestricted) form of (C.=) leads to theoretical difficulties in connection with our concept of individual. (As Quine has insisted, Leibniz's law must hold for individuals – otherwise there is something wrong with the very notion.) However, I shall suggest later in this paper that the justified and true element in Quine's claim can be caught in a different way without modifying at all what has been said so far.

V. A PARADOX INDEPENDENT OF IDENTITY

It is readily seen that some changes are needed in (C.E) and (C.U) to avoid paradoxical applications similar to those that arise in connection with the strong form of (C.=). The following implication is a case in point:

$$(1) \qquad N(a=a) \supset (Ex)N(x=a).$$

Reading 'a' = 'the next president of the United States', the antecedent of (1) says that necessarily the next president is the next president, which is obviously true. The consequent says that there is someone who necessarily is the next president, i.e. whose election is inevitable. On any reasonable interpretation of necessity, this is false. Hence (1) is (contingently) false.

Yet it would be logically true if our definitions were accepted unmodified. This may be seen as follows: Assume that (1) is not logically true, i.e. assume that there is a model system Ω and a model set $\mu \in \Omega$ such that

(2) '$N(a=a)$' $\in \mu \in \Omega$ counter-assumption
(3) '$(Ux)M(x \neq a)$' $\in \mu \in \Omega$ counter-assumption.

Then it follows:

(4) '$M(a \neq a)$' $\in \mu$ from (3) by (C.U).

Here (4) and (2) violate the stronger form of (C.$\sim$), showing the impossibility of (2)–(3) and hence the logical truth of (1). If the stronger form is not available, we can argue as follows:

(5) '$(a \neq a)$' $\in \lambda$ for some alternative $\lambda \in \Omega$ to μ, by (C.M*).

Here (5) violates (C.self$\neq$), yielding the same conclusion.

Extremely simple though this argument is, it deserves several comments. Its very simplicity shows – or at least very strongly suggests – that the only possible source of trouble here is the condition (C.U). The other conditions relied on were (C.$\sim$) (or alternatively (C.self$\neq$)) and (C.M*), none of which is subject to reasonable doubts in this context. (Furthermore, denying the possibility of our preliminary simplifications presupposes giving up classical logic in a fairly radical way.) Moreover, contrary to many suggestions, the trouble cannot be blamed on the failure of the substitutivity of identity in modal contects, for (C.=) was not used in the argument at all. Nor can the paradoxical result be laid to the possibility of illicitly 'importing' new individuals to model sets – a phenomenon known to cause trouble elsewhere.[8]

Our example is similar to Quine's favorite brand of illustrations. Instead of 'proving' (1), we likewise could have 'proved'

'N (the number of planets = the number of planets) $\supset$
$(Ex)N$(the number of planets $= x$)'.

The fact that the fallaciousness of our proof does not turn at all on our assumptions concerning identity belies (it seems to me) Quine's (and Føllesdal's) emphasis on the failure of the substitutivity of identity as the source of trouble in this area.[9]

Doubts are perhaps raised concerning my claim that (1) is false. It seems to me obvious, however, that on the intended interpretation of (1) it *is* false, and that these doubts can be traced to what we are inclined to say of the translation of (1) into more or less ordinary discourse and of the logical behavior of this translation. What is being presupposed by the above conditions is clearly an interpretation of '*M*' such that *Mp* is true in a possible world iff *p* were true in at least one alternative possible world. If we understand the notion of alternativeness, we can understand this independently of how we are inclined to express ourselves in ordinary language. Here, we do not even need more than a rudimentary understanding of what is involved, for all that is presupposed by the truth of the antecedent of (1) is that there is no possible world in which it is true to assert '$a \neq a$'. And this much is surely completely uncontroversial.

The same line of thought suggests that we cannot very well doubt the applicability of (C.self$\neq$) on the intended interpretation.

I am not saying that it is impossible to try to interpret modal operators in some other way, at least *ad hoc*. However, no interpretation essentially different from mine has ever been given a satisfactory semantic development. The onus is thus on those who want to understand the antecedent of (1) in some way different from mine. Moreover, even if an alternative interpretation of (1) (or perhaps rather of its translation into an 'ordinary language') can be given, I am sure it will turn out to be analyzable in terms of the interpretation I am presupposing. (Later, I shall in fact indicate one such interpretation of the antecedent of an ordinary-language statement reminiscent of (1).) And even if there are irreducible competing readings, I can have my paradox (counter-example to the above conditions) simply by restricting my attention to the intended reading of (1).

In case you are here worried about the possible failure of *a* to exist, you are welcome to replace (1) by

(1*) '$((Ex)(x = a) \,\&\, N(a = a)) \supset (Ex)N(x = a)$'

which is as paradoxical as (1) and which can be shown to be logically true in the same way as (1) even when existential presuppositions are eliminated. The only additional assumption we have to make here is that there is such a person as the next president of the United States, i.e., that the United States will continue as a democracy – a factual assumption which I hope we are entitled to make.

Hence the truth of the antecedent of (1) on the intended interpretation is beyond doubt. The same is the case with the falsity of the consequent, again given the intended interpretation. In this case, the relevant aspect of the interpretation is the idea that bound (bindable) variables take ordinary individuals (e.g. persons) as their values. I hope that I do not have to try to justify this assumption here after everything Quine and Føllesdal have said concerning its indispensability.

Instead of rehearsing their weighty and eloquent reasons I shall merely register the fact that this assumption amply justifies my claim that the consequent of (1) is false. For who among the actual candidates is the invincible one? The labors of the supporters of the several candidates suggest that they at least are not convinced of the historical (and even less of the logical) necessity of their favorite's being nominated and elected.

The underlying semantical reason why the existential generalization (1) (or (1*) fails is also obvious. Under the different courses of events that are in fact possible the term 'the next president of the U.S.' refers to different politicians. Hence we cannot go from a statement, however true, about these different individuals to a statement which says that there is some one (unique) individual of which the same is true.

VI. AN AMBIGUITY IN ORDINARY LANGUAGE

Moreover, in addition to justifying the falsity of the consequent, Quine's assumption perhaps also helps to remove a residual awkwardness about the truth of the antecedent of (1). I am not denying that such ordinary-language statements as

(6) 'the next president of the United States is necessarily the next president of the United States'

can occasionally be understood so as to be false.

In general, my basic semantical idea of 'possible worlds' shows that almost any ordinary-language statement in which a singular term occurs within a modal context is in principle potentially ambiguous. Such a statement can sometimes be understood in (at least) two different ways. It can be taken to be about the different individuals which the term picks out in the different possible worlds that the modal operator invites us to consider. However, often it can also be understood as being about the unique individual to which the term in fact refers (i.e. refers in the actual world). More specifically, one faces this choice of interpretations at each occurrence of the term. For instance, the two interpretations of the first occurrence of 'the next president of the U.S.' in (6) yield

(7) 'necessarily whoever is the next president of the U.S. is identical with whoever is the next president of the U.S.'

and

(8) 'the man who in fact will be the next president of the U.S. is necessarily identical with the next president of the U.S.',

respectively. The former (7) is what we took '$N(a = a)$' to express. Because bound variables range over actual individuals, the latter (8) can be formalized by

(8*) '$(Ex)\,(x = a \;\&\; N(x = a))$'

or perhaps rather by

(8**) '$(x)\,(x = a \supset N(x = a))$'.

Unlike (7), (8)–(8**) can certainly be false – and in fact seem to be so. If you felt uncomfortable about my bland initial assertion that the antecedent of (1) is true, you had some right to do so, for the best translation of the antecedent into ordinary language admits of an interpretation which makes it false. This does not belie my point, however, that under the intended interpretation (1) is in fact false. On the contrary, our observations dispose of a plausible (but false) reason for doubting my suggestion.

VII. MODIFYING THE QUANTIFIER CONDITIONS

The main thing we have established by all of this is that the condition (C.U) has to be modified. But how? It is here that my principal working hypothesis comes in. The only essential assumption I shall make is that somehow or other we can modify (or amplify, if need be) our language so that the desired condition on which (C.U) can be restored can be expressed in the language itself. (Cf. the methodological precepts mentioned toward the end of 'Existential Presuppositions and Their Elimination' in the present volume, pp. 42–43.)

This is not quite enough, however, for we must be able to allow for slightly different conditions depending on how 'x' occurs in p. Here our semantical point of view yields useful hints. Obviously, whatever goes wrong with (1) is due to the fact that under different courses of events we consider possible 'a' refers to different individuals. To forestall this, it must be required as an additional premise that 'a' does not exhibit this kind of referential multiplicity. Likewise, a further clause is needed in the requirements (C.U) which was used in the 'proof' of (1). What this additional condition must express is here the uniqueness of the reference of '*b*' in all the different possible worlds as a member of which we are considering '*x*' in *p*.

This is determined by the number of modal operators within the scope of which the variable '*x*' occurs at its different appearances in *p*. Let us assume that these numbers are $n_1, n_2, \ldots$. This fact will in the sequel be expressed by saying that the *modal profile* of *p* with respect to '*x*' is $n_1, n_2 \ldots$. Let us assume that in these circumstances some (so far completely unspecified) formula

(9) '$Q^{n_1, n_2, \ldots}(b)$'

expresses the desired condition. The only thing we are assuming about (9) is that the only free singular term in it is *b*. Then it is obvious how the crucial condition (C.U) has to be reformulated:

(C.U$_q$) If $(Ux)p \in \mu \in \Omega$, if the modal profile of *p* with respect to '*x*' is $n_1, n_2, \ldots$ and if '*b*' occurs in the formulas of some member of Ω, then $[p(b/x) \vee \sim Q^{n_1, n_2, \ldots}(b)] \in \mu$.[10]

Here '*b*' is to all intents and purposes a completely arbitrary free singular term. The requirement that it appears in the members of some $\lambda \in \Omega$ is inserted simply to avoid having to speak of an unlimited class of free singular terms.

It is also clear that we have to change the dual condition (C.E) likewise. From the truth of an existentially quantified statement, more follows than the truth of some substitution-instance. A substitution-instance with respect to a term with the right sort of unique reference can always be introduced *salva* consistency:

(C.E$_q$) If $(Ex)p \in \mu$, and if the modal profile of *p* with respect to '*x*' is $n_1, n_2, \ldots$ then $p(a/x) \in \mu$ and '$Q^{n_1, n_2, \ldots}(a)$' $\in \mu$ for some '*a*'.

VIII. THE MODIFICATION OF QUANTIFIER CONDITIONS IS UNIQUELY DETERMINED

The remarkable thing is that (C.E$_q$)–(C.U$_q$) suffice, together with the rest of our conditions and conventions, to determine what the logical power of '$Q^{n_1, n_2, \cdots}(b)$' is in the sense of showing how it can be expressed in our original symbolism. Given one further assumption we can show that the following formula is logically true:[11]

$$(10) \qquad \text{'}(Ex)\,[N^{n_1}(x = b)\ \&\ N^{n_2}(x = b)\ \&\ \cdots] \equiv Q^{n_1, n_2, \cdots}(b)\text{'}.$$

In order to prove this, it suffices to prove implications in both directions. In order to prove such an implication, it suffices to reduce *ad absurdum* the assumption that its antecedent and the denial of its consequent are satisfiable together, i.e., can occur in the same member μ of a modal system Ω. This can be done as follows:

Left to right: Assume (counter-assumption)

$$(11) \qquad \text{'}(Ex)\,[N^{n_1}(x = b)\ \&\ N^{n_2}(x = b)\ \&\cdots]\text{'} \in \mu \in \Omega$$

and

$$(12) \qquad \text{'}{\sim}\, Q^{n_1, n_2, \cdots}(b)\text{'} \in \mu .$$

Then by (C.E$_q$) we have for some 'a'

$$(13) \qquad \text{'}[N^{n_1}(a = b)\ \&\ N^{n_2}(a = b)\ \&\cdots]\text{'} \in \mu$$

$$(14) \qquad \text{'}Q^{n_1, n_2, \cdots}(a)\text{'} \in \mu .$$

Furthermore, we have from (13) by (C.&)

$$(15) \qquad \text{'}N^{n_1}(a = b)\text{'} \in \mu$$

$$\text{'}N^{n_2}(a = b)\text{'} \in \mu$$

$$---$$

Now at this stage we have to make some assumptions in order to get anywhere. We have to assume that with respect to the substitutivity of identity '$Q^{n_1, n_2, \cdots}(b)$' behaves in the same way as a formula whose modal profile with respect to 'b' is n_1, n_2,... It is very hard to see how this could fail to be the case, for '$Q^{n_1, n_2, \cdots}(b)$' is intended to guarantee the uniqueness of reference of the term 'b' precisely in the same possible worlds as a member of which a sentence p speaks of b when the modal profile of p with respect to 'b' is $n_1, n_2, \ldots$ In short, as far as b is concerned, '$Q^{n_1, n_2, \cdots}(b)$' speaks of the same possible worlds as p and might therefore be expected to behave vis-à-vis substitution in the same way as a sentence with the same profile as p. What this assumption means is that (C.N=) is applicable to (14) and (15), yielding

$$(16) \qquad \text{'}Q^{n_1, n_2, \cdots}(b)\text{'} \in \mu .$$

Assuming the stronger form of (C.$\sim$), this contradicts (12) and therefore proves the desired implication.

Right to left: Assume (by way of counter-example) that

(17) $`Q^{n_1, n_2, \cdots}(b)\text{'} \in \mu \in \Omega$

and that

(18) $`(Ux)\,[M^{n_1}(x \neq b) \vee M^{n_2}(x \neq b) \vee \cdots]\text{'} \in \mu.$

Then we have from (17)–(18) by (C.U$_q$)

(19) $`\sim Q^{n_1, n_2, \cdots}(b) \vee [M^{n_1}(b \neq b) \vee M^{n_2}(b \neq b) \vee \cdots]\text{'} \in \mu$

Hence by (C.$\vee$) we have either '$\sim Q^{n_1, n_2, \cdots}(b)$'$\in \mu$ which is excluded by (17) and the stronger form of (C.$\sim$), or else

(20) $`M^{n_i}(b \neq b)\text{'} \in \mu$

for some i. From (20) we have by n_i applications of (C.M*)

(21) $`(b \neq b)\text{'} \in \lambda$

for some alternative λ to μ (n_i times removed from μ). But (21) violates (C.self $\neq$), showing the impossibility of (17)–(18), thus proving the desired implication, and thereby demonstrating the logical truth of (10).

Thus there is no need to introduce any new symbolism. The auxiliary condition that guarantees uniqueness can be expressed in our original notation. Nor is there any choice as to what the interrelation of the auxiliary condition to other formulas is. In fact, we can simply replace (C.U$_q$) and (C.E$_q$) by the following conditions:[12]

(C.U$_1$) If $(Ux)p \in \mu \in \Omega$, if the modal profile of p with respect to 'x' is $n_1, n_2, \ldots$, and if 'b' occurs in the formulas of some member of Ω, then '$\sim (Ex)(N^{n_1}(x=b) \,\&\, N^{n_2}(x=b) \,\&\, \cdots) \vee p(b/x)$'$\in \mu$.

(C.E$_1$) If $(Ex)p \in \mu$, and if the modal profile of p with respect to 'x' is $n_1, n_2, \ldots$, then, for some 'a', $p(a/x) \in \mu$ and '$(Ex)\,(N^{n_1}(x=a) \,\&\, N^{n_2}(x=a) \,\&\, \ldots)$'$\in \mu$.

It is understood that the order of conjuncts does not matter here. Moreover, we can require in (C.E$_1$) also that every formula obtained from

(*) $`(Ex)\,(N^{n_1}(x=a) \,\&\, N^{n_2}(x=a) \,\&\, \cdots]\text{'}$

by omitting some of the conjuncts is also in μ. In fact, we shall require in general that these shorter formulas are in μ if (*) is in it.

We have seen that, given the assumptions indicated above, the conditions (C.U_1) and (C.E_1) are essentially the only way of reconciling quantification into modal contexts with the usual semantical conditions for logical constants. The solution which they present to the problem of quantifying into modal contexts can also be motivated directly in intuitive terms. Quine's criticism of quantified modal logic is predicated on the idea that quantifiers range over genuine, well-defined individuals. Now a free singular term (say '*a*') which picks out different individuals in the different possible worlds one is considering cannot specify such a well-defined individual. In order for '*a*' to specify one, there must be some *one and the same* individual to which it refers in all the possible worlds one must take into account. But this is just what (*) expresses in the case in which the relevant possible worlds are those $n_1, n_2, \ldots$ steps removed from the one μ describes. Hence what (C.U_1) and (C.E_1) imply may be partially expressed by saying that according to them a singular term is an acceptable substitution-value for a bound variable if and only if it picks out one and the same individual in all the relevant possible worlds. These possible worlds are of course precisely the ones as members of which one considered the values of the bound variable '*x*' in the quantified sentences $(Ex)p$ and $(Ux)p$ mentioned in (C.U_1) and (C.E_1). These conditions thus say just what one can expect on the basis of the Quinean interpretation of quantification in the first place.[13]

It is especially important to appreciate this semantical situation in view of the widespread misinterpretations of the intended role of our uniqueness premises (22) or (24). Their function has frequently been taken to be to restrict somehow the range of individuals over which one's bound variables range. This is a serious (and, it seems to me, unprovoked) oversimplification. (This misinterpretation has been perpetrated, among others, by Hector-Neri Castañeda and Wilfrid Sellars.) When we have to consider our individuals as members of several possible worlds, the whole notion of 'ranging over' becomes so oversimplified as to be of little explanatory value, and we cannot in any case describe satisfactorily the role of the uniqueness premises in terms of this notion. A restriction on the range of one's bound variables restricts them to some subset of actually existing individuals. What our quantifier conditions (C.E_1) and

$(C.U_1)$ involve is not this sort of restricted quantification, but a recognition of the fact that we have to consider those individuals over which bound variables 'range' as members of several possible worlds. This necessitates spelling out the fact that in considering one particular 'value' of a bound variable we must consider *one and the same* individual in all the relevant possible worlds. This is what $(C.E_1)$ and $(C.U_1)$ accomplish, and it is completely obvious that this task cannot be performed by ordinary relativization of quantifiers ranging over actual objects.

IX. GENERALIZATIONS

These considerations can be extended to the case in which we have any number of pairs of modal operators '$N^{(i)}$', '$M^{(i)}$'. We shall assume that they are distinguished from each other by superscripts. Furthermore, these operators may be relativized to an individual or a set of individuals, to be indicated by a subscript. (It is convenient to assume that one of the available subscripts stands for a 'null individual' which characterizes an unrelativized modality). Semantically speaking, each new subscript brings in a new alternativeness relation. (These different relations may of course be interrelated in different ways.) For the time being we assume that no bound variables occur as subscripts to modal operators.

The occurrence of free singular terms as subscripts of modal operators necessitates a dual extension of (C.=) to cover this case:

(C.=N) If '$(a = b)$' $\in \mu$ and $N_a p \in \mu$, then $N_b p \in \mu$.
(C.=M) If '$(a = b)$' $\in \mu$ and if $M_a p \in \mu$, then $M_b p \in \mu$.

Here 'N', 'M' are assumed to be any modal operators to which subscripts can be attached.

Let us now consider an occurrence of 'x' in p. If the modal operators (in order) within the scope of which 'x' occurs at this place are characterized by the subscripts and superscripts ${}^{(i_1)}_{a_1}, {}^{(i_2)}_{a_2}, \ldots$, then this list is said to indicate the *modal character* of the occurrence in question. (Notice that the subscript of an operator is thought of as being *outside* the scope of that operator.) A sequence of the modal characters of all the occurrences of 'x' in p will be called (by modifying our earlier definition somewhat) the *modal profile* of p with respect to 'x'.

It is easy to see how the above conditions $(C.U_1)$ and $(C.E_1)$ are to be modified so as to apply to this general case. The only change needed concerns the auxiliary formula that takes over the role of

$$(22) \qquad (Ex)\,(N^{n_1}(x=b)\ \&\ N^{n_2}(x=b)\ \&\cdots).$$

If the modal profile of p with respect to 'x' is

$$(23) \qquad {}^{(i_{11})}_{a_{11}}\,{}^{(i_{12})}_{a_{12}}\cdots,\ {}^{(i_{21})}_{a_{21}}\,{}^{(i_{22})}_{a_{22}}\cdots,\ \cdots$$

then the role of (22) will be played by the formula

$$(24) \qquad \text{`}(Ex)\,[N^{(i_{11})}_{a_{11}}\,N^{(i_{12})}_{a_{12}}\cdots(x=b)\ \&\ N^{(i_{21})}_{a_{21}}\,N^{(i_{22})}_{a_{22}}\cdots(x=b)\ \&\cdots]\text{'}.$$

Otherwise, the resulting conditions ($C.U_m$) and ($C.E_m$) will be like ($C.U_1$) and ($C.E_1$).

If the alternativeness relation that goes together with a pair of modal operators – say '$N^{(i)}$', '$M^{(i)}$' – is transitive, then in (24) repetitions of '$N^{(i)}$' (with the same subscript, if any) can be disregarded. Interrelations between different modal operators (or with the same modal operator with different subscripts) have to be studied *in casu*.

Apart from this qualification, our formulations above seem to cover almost everything that is needed for a satisfactory treatment of modal logics, including modal logics with several different kinds of modal operators.

X. EXISTENTIAL PRESUPPOSITIONS AS SPECIAL CASES OF UNIQUENESS PRESUPPOSITIONS

Several comments are in order here which may elucidate further the import of what I have said.

In the formulations ($C.U_1$) and ($C.E_1$), the case $n_i=0$ was included. Then N^{n_i} will be an empty sequence of N's, and (22) will contain a conjunct of the form '$(x=b)$'. The corresponding possibility that the modal character of some of the occurrences of 'x' in p in the empty sequence is likewise assumed to be needed, yielding a conjunct of the form '$(x=b)$' in (24). In case no modal operators are around at all, these are all the cases we have to worry about. Then the crucial clauses (22) and (24) which serve to guard us against failures of uniqueness reduce to clauses of the form '$(Ex)\,(x=b)$' which express the familiar existential presuppositions. The necessity of formulating the presuppositions precisely in this way emerges as a special case of the argument given above. This brings out what I meant in the beginning of this paper when I emphasized that what we need in modal logic is an elimination of uniqueness presuppositions analogous to the elimination of existential presuppositions. The failure of our usual logical laws in modal contexts, signalled by the invalidity of (1), can be traced to the fact that not all our singular terms will exhibit the right kind of uniqueness of reference when different possible worlds are compared with each other. Hence in our basic conditions for existence and universality we cannot assume such uniqueness, which has to be postulated explicitly when it holds. We have seen that these uniqueness presuppositions and existence presuppositions can both be uncovered by one and the same argument.

XI. FURTHER OBSERVATIONS. THE PARITY OF IDENTICAL INDIVIDUALS

Among other things, we can now also see what went wrong in our 'proof' of (1). In order to get (4) from (2)–(3), we would need an additional assumption

(25) $`(Ex)N(x = a)' \in \mu$

which not only is false but which would contradict (3) in virtue of (C. $\sim$). The 'proof' of (1) is thus strictly a *petitio principii.*

These observations give us a considerable chunk of modal logic in a semi-semantical formulation. It is easily seen, among other things, that none of the famous critical formulas will be logically true whose truth depends on 'moving individuals from one possible world to another', i.e. depends on assuming that whenever an individual exists in one world, it exists in certain others.[14] It is easy to construct a counter-example e.g. to the Barcan formula and to formulas of the form

(26) $(Ex)Np \supset N(Ex)p$

where we may for simplicity assume that p is atomic.[15] Hence the annoying necessity of having to modify (C.N$^+$) and perhaps even *modus ponens* is automatically avoided.[16]

There is a further point, however, that deserves to be made. We have discovered that whenever (in the presence of a single pair of modal operators) the modal profile of p with respect to 'x' is $n_1, n_2, \ldots$, then

(27) $\quad$ '$(Ex)\,(N^{n_1}(x = b)\ \&\ N^{n_2}(x = b)\ \&\ \ldots)$'

suffices to restore existential generalization with respect to 'b' in $p(b/x)$ so as to yield $(Ex)p$. In fact, the force of (27) could almost be expressed by reading it 'b is an individual' (i.e. for the purposes of a context with the same modal profile as p). But if this restoration of the reference of 'b' to the status of a 'real' individual is to succeed, it might be suggested, then the same auxiliary premises should restore the substitutivity of identity, too. For the same things should really be said of identical individuals. This suggests adopting the following condition:

(C.ind=) If $p \in \mu$, '$(a = b)$' $\in \mu$, and if q results from p by interchanging 'a' and 'b' in a number of places which are within the scope of $n_1, n_2, \ldots$ modal operators, respectively, and if

'$(Ex)\,[(x = a)\ \&\ N^{n_1}(x = a)\ \&\ N^{n_2}(x = a)\ \&\cdots]$' $\in \mu$
'$(Ex)\,[(x = b)\ \&\ N^{n_1}(x = b)\ \&\ N^{n_2}(x = b)\ \&\cdots]$' $\in \mu$

then $q \in \mu$.

Here the displayed existentially quantified formulas may be replaced by some of their admissible variants, i.e. by formulas obtained from them by trading 'x' for some other bound variable and/or changing the order of some identities and conjunctions. In view of (C.N=), (C.ind=) is equivalent to the following condition:

(C.ind$=_0$) If '$(a = b)$' $\in \mu$,
'$(Ex)\,[(x = a)\ \&\ N^{n_1}(x = a)\ \&\ N^{n_2}(x = a)\ \&\cdots]$' $\in \mu$,
'$(Ex)\,[(x = b)\ \&\ N^{n_1}(x = b)\ \&\ N^{n_2}(x = b)\ \&\cdots]$' $\in \mu$,
then '$[N^{n_1}(a = b)\ \&\ N^{n_2}(a = b)\ \&\cdots]$' $\in \mu$.

(Here the existentially quantified formulas may again be replaced by suitable admissible variants of theirs.)

The reason why the first conjuncts '$(x = a)$' and '$(x = b)$' are needed in (C.ind=) and (C.ind$=_0$) is obvious. In order for the 'actual' identity '$(a = b)$' to have any effects, the references of 'a' and 'b' have to be 'genuine individuals' also in so far as the world described by μ is concerned.

Apparently for many modalities (C.ind=) is an acceptable condition.

(There are rather plausible-looking counter-examples to it in epistemic logic. I believe that I can nevertheless explain them away. To attempt to do it here would take us too far, however.[17]) The way in which it can be generalized so as to apply when several pairs of modal operators (with or without subscripts) are present should be obvious on the basis of our earlier discussion.

XII. QUINE VINDICATED (?)

It seems to me that (C.ind=) really brings out the true element in Quine's emphasis on the substitutivity of identity as a test of the normality of our interpretation of the concept of individual. The true element, I submit, is the parity of identical *individuals*. Not any two singular terms which pick out the same individual in the actual world are intersubstitutable in modal contexts, for they may refer to different individuals in other possible worlds we have to consider. However, whatever is said of a genuine (unique) *individual* can always be said of another individual identical with it. This is precisely what (C.ind=) spells out when the notion of an individual is relativized to a particular context (class of possible worlds).[18]

By the same token, 'genuine individuals' in the sense just indicated must exhibit other kinds of nice predictable behavior. One fairly obvious requirement of this sort is the following:

(C.ind=E) If '$(a=b)$' $\in \mu$

$$\text{'}(Ex)\,[(x=a)\ \&\ N^{n_1}(x=a)\ \&\ N^{n_2}(x=a)\ \&\cdots]\text{'} \in \mu$$
$$\text{'}(Ex)\,[(x=b)\ \&\ N^{m_1}(x=b)\ \&\ N^{m_2}(x=b)\ \&\cdots]\text{'} \in \mu$$

then

$$\text{'}(Ex)\,[(x=a)\ \&\ (x=b)\ \&\ N^{n_1}(x=x)\ \&\ N^{n_2}(x=x)\ \&\cdots\ \&\ N^{m_1}(x=x)\ \&\ N^{m_2}(x=x)\ \&\cdots]\text{'} \in \mu .$$

Instead of the first two existentially quantified formulas, we may here have any admissible variants of theirs.

The intuitive motivation of (C.ind=E) will be commented on later. Meanwhile, it may be pointed out that together (C.ind=) and (C.ind=E) seem to catch Quine's intentions very well. By their means the validity of all formulas of the following form can be demonstrated:

$$(28) \qquad (Ux)\,(Uy)\,((x=y) \supset (p \supset q))$$

where p and q are like each other except for an interchange of 'x' and 'y' at a number of places. Now it is in terms of these formulas that Quine frequently formulates his point about the substitutivity of identity. Moreover, the reasons Quine actually gives for the substitutivity principle are admirably suited to motivate the adoption of a principle of parity for identical *individuals*, whereas I do not see that they carry any weight whatsoever as a defense of the substitutivity of *de facto* coreferential free singular terms, i.e. as a defense of the unqualified form of (C.=). One general defect of Quine's and Føllesdal's discussions of the substitutivity principle seems to be a failure to emphasize sufficiently the distinction between the different variants of the principle.

In order to see that the bound-variable version of the substitutivity principle (28) is valid, we may argue as follows:

(29) $(Ex)(Ey)((x = y) \& p \& \sim q) \in \mu \in \Omega.$

This is the counter-assumption (for some model set μ and modal system Ω). It can be reduced *ad absurdum* as follows:

(30) $(Ey)((a = y) \& p(a/x) \& \sim q(a/x)) \in \mu$

(31) $`(Ex)[(x = a) \& N^{n_1}(x = a) \& N^{n_2}(x = a) \& \cdots]' \in \mu$

(32) $(a = b) \& p(a/x)(b/y) \& \sim q(a/x)(b/y) \in \mu$

(33) $`(Ex)[(x = b) \& N^{m_1}(x = b) \& N^{m_2}(x = b) \& \cdots]' \in \mu.$

Here $n_1, n_2, \ldots$ is the modal profile of p with respect to 'x' and $m_1, m_2, \ldots$ the modal profile of q with respect to 'y'. Of these steps, (30)–(31) follow from (29) by (C.E$_1$), and (32)–(33) follow from (30) likewise.

From (32) we have by (C.&)

(34) $`(a = b)' \in \mu$

(35) $p(a/x)(b/y) \in \mu$

(36) $\sim q(a/x)(b/y) \in \mu.$

From (31), (33) and (34) we obtain by (C.ind=E)

(37) $`(Ex)[(x = a) \& (x = b) \& N^{n_1}(x = x) \& N^{n_2}(x = x) \& \cdots \& N^{m_1}(x = x) \& N^{m_2}(x = x) \& \cdots]' \in \mu.$

Hence we have by ($C.E_1$) and (C.&) for some 'd'

(38) '$(d = a)$' $\in \mu$

(39) '$(d = b)$' $\in \mu$

(40) '$(Ex)\,[(x = d)\ \&\ N^{n_1}(x = d)\ \&\ N^{n_2}(x = d)\ \&\cdots\&$
$N^{m_1}(x = d)\ \&\ N^{m_2}(x = d)\ \&\cdots]$' $\in \mu$

and therefore *a fortiori*

(41) '$(Ex)\,[(x = d)\ \&\ N^{n_1}(x = d)\ \&\ N^{n_2}(x = d)\ \&\cdots]$' $\in \mu$

(42) '$(Ex)\,[(x = d)\ \&\ N^{m_2}(x = d)\ \&\ N^{m_2}(x = d)\ \&\cdots]$' $\in \mu$.

By (C.ind=) we now have from (35), (31), and (41)

(43) $p(d/x)(b/y) \in \mu$

and in the same way from (43), (33), and (42)

(44) $p(d/x)(d/y) \in \mu$.

By the same line of argument but starting from (36) instead of (35) we have

(45) $\sim q(d/x)(d/y) \in \mu$.

But $p(d/x)(d/y)$ and $q(d/x)(d/y)$ are by assumption identical. Hence (44) and (45) violate the stronger form of (C.$\sim$). This contradiction completes our reductive argument and hence establishes the desired validity.

It is seen at the same time, however, that Quine's emphasis on the substitutivity of identity as the main test that our concept of an individual is all right may not have been entirely happy, even when his point is interpreted in the way we just did. Quine is absolutely right in insisting that the only way of carrying out the normal, intended interpretation of quantification is to require that bound variables range over genuine individuals. What this leads us to, however, is primarily a modification of our conditions on quantifiers, and only secondarily an addition to our conditions on identities. (In fact, we have seen that at least some of the paradoxes that otherwise ensue arise independently of our ways with identity.) The conditions (C.ind=) and (C.ind=E) which serve to satisfy Quine's requirement concerning the substitutivity of identity embody happy afterthoughts rather than indispensable elements of our treatment of identity.

What we have found contains the essential features of a general theory of modality – both for the case of a single pair of modal operators and for the general case of any number of pairs of operators. What is missing is (*inter alia*) a treatment of the case in which bound variables occur as subscripts. No such treatment will be attempted here. A discerning reader may perhaps already perceive what form it can take when conducted e.g. in terms of the ε-technique of Hilbert.

Another topic that will largely be left untouched here is the question of the special assumptions which can be made concerning different particular modalities (e.g. various assumptions concerning the properties of their alternativeness relations, such as transitivity, reflexivity absolutely or under certain conditions, symmetry, etc.) They will have to be dealt with in various special theories of particular modal notions, it seems to me.

XIII. EPISTEMIC LOGIC

Among other things, we obtain in this way a formulation of epistemic logic which is in some respects a modified and extended version of the system presented in my book *Knowledge and Belief.*

The main additional assumption we may want to make here is (I have argued) the transitivity of the epistemic alternativeness relation. Or, rather, this assumption characterizes philosophers' strong sense of knowledge in which it is contrasted to merely possessing true information. In the presence of just one pair of epistemic notions K_a, P_a (corresponding to 'N' and 'M', and expressing what the bearer of 'a' knows and his 'epistemic possibility', respectively) the only possible types of auxiliary premises (uniqueness and/or existence presuppositions) will then be of the following kinds

(46) '$(Ex)\,(x = b\ \&\ K_a(x = b))$'

(47) '$(Ex)\,K_a(x = b)$'

(48) '$(Ex)\,(x = b)$'

depending on whether we are considering a formula in which 'b' occurs both inside and outside the scopes of 'K_a' and/or 'P_a', only inside, or only outside.

It is important to realize that these three conditions are – or at least can be assumed to be – logically independent apart from the fact that (46) logically implies the other two. (This implication follows easily from the assumptions made earlier.) The independence shows that, although existence presuppositions and uniqueness presuppositions are largely parallel to each other and although they can be discussed essentially in the same way, they are nevertheless materially different assumptions. The cases in which the implication from (47) to (48) fails are likely to be somewhat marginal, but I do not see any persuasive reasons why they should be ruled out. They will amount to cases where *a* knows who (or what) someone (or something is, 'should he (or it) exist', although it so happens that he (it) does not. Allowing for such cases seems a natural course in view of certain puzzling examples, though it does not seem to be an indispensable way out of these difficulties.

Likewise, disallowing the implication from (47) and (48) together to (46) would not lead to a violation of any other assumptions we are likely to make concerning the notion of knowledge. However, the 'success grammar' of knowing makes this implication a very natural assumption.

The independence of the analogues to (46)–(48) is of course completely obvious in the case of belief instead of knowledge.

Part of the force of the transitivity assumptions which characterizes the strong sense of knowledge can be caught by the following 'transfer assumption':

(C.EK=EK=*) If '$(Ex)\, K_a(x=b)$'$\in\mu\in\Omega$ and if $\lambda\in\Omega$ is an epistemic alternative to μ with respect to '*a*', then '$(Ex)\, K_a(x=b)$'$\in\lambda$.

XIV. MODELS FOR MODEL SYSTEMS

So far I have not said anything about the kind of 'real' semantics that might go together with my semi-semantical treatment. I suspect that in the study of modal notions, a treatment of their logic by reference to model sets and model systems is in fact simpler and more straightforward than a treatment in ordinary semantical terms. Nevertheless, a few words about what a genuine semantics for my modal logics will look like might clarify the situation and also clarify the relation of my approach to that used by other logicians.[19]

My treatment involves, first of all, an innocuous assumption that we have a name available for each individual and for each entity of any other type that we want to consider. I shall not pause to rehearse the standard objections to this simplifying assumption or the equally standard rejoinders that can be made to these objections. A similar procedure is in any case familiar enough in ordinary non-modal logic.

The main effect of this simplifying assumption is that we can in most cases formulate the truth-conditions of sentences of different kinds very easily in terms of truth-conditions for certain simpler sentences. This is essentially what happens in the conditions defining a model set and a model system. The reason why we need a suitable supply of singular terms here is that the simpler sentences just mentioned are often substitution-instances of the original ones with respect to certain particular kinds of singular terms (constants).

There are of course types of sentences whose truth-conditions cannot be reduced further in this way. This is the case with atomic sentences and identities. Discarding existential presuppositions adds a new class of such irreducible sentences, viz. the Quinean sentences of the form (48). (I shall not discuss here what kind of semantics is appropriate to them.) Now what happens when uniqueness presuppositions are given up – or, rather, replaced by explicitly formulated uniqueness premises – is that a further class of such irreducible sentences is created. These are precisely the sentences (formulas) of form (22) (or, more generally, of form (24)). Thus the main question which my treatment of modal logics leaves without a sufficiently explicit discussion is the question as to what the truth of these sentences 'really' amounts to.

A partial answer is nevertheless implicit in the above discussion. The intuitive idea on which this discussion is based is the following. Each free singular term picks out a member (an individual or perhaps rather a particular 'stage' or 'manifestation' of an individual) from each possible world we are considering. (I am disregarding the possible emptiness of singular terms here, if only in order to simplify my discussion.) However, the individuals so picked out need not be identical (i.e. they need not be 'manifestations' of the same individual in all these worlds). Only some free singular terms always pick out the same individual. They are the ones that satisfy the appropriate uniqueness conditions (22) (or (24)). In order for this to be an objectively defined notion and in order to speak of the totality of individuals which can in this way manifest themselves in different possible worlds, we must assume that we are given a particular objectively determined set of functions each of which picks from all the appropriate possible worlds the manifestation of *one and the same* individual. These functions, in a sense, are thus the real individuals we are talking about in our sentences, while the members of the several possible worlds are better thought of as so many roles that those individuals may play. In formulating truth-conditions for sentences in which we quantify over individuals, we must speak of the existence of suitable 'individuating functions' of the kind just mentioned. However, if we assume that for each such function there exists a singular term picking out just the several values of this function from those possible worlds we are considering, then we can in fact formulate the truth-conditions by reference to the existence of such singular terms. These exceptionally well-behaved terms will be characterized by the fact that they satisfy the appropriate uniqueness conditions (22) or (24). Thus we are inevitably led to the precise conditions ($C.U_m$) and ($C.E_m$) given above, re-interpreted as truth-conditions for quantified sentences in modal contexts. In this way we can see what kind of semantics goes together with my semi-semantical treatment of modality. It may be that my conditions can be viewed as rules of disproof rather than semantical conditions proper. However, it is very easy to see what semantical counterparts they have.

It must be admitted, however, that for some other purposes a usual semantical approach is more straightforward. Although conditions on model sets and model systems are usually obvious (when acceptable), it is not always equally clear that we have exhausted by their means all the assumptions we have to make in this area. In fact, a moment's reflexion shows that the conditions so far recorded are yet insufficient to capture all the semantical principles we are trying to codify. Apart from the treatment of bound individual variables as subscripts to modal operators, there are at least two kinds of assumptions that remain to be made. One of them is obvious, and merely brings out the possibility of bearing one and the same trans world lines of cross-identity from the point of view of different model sets:

(C. ind*) If

(*) $$\text{`}(Ex)\,[N^{n_1}(x=b)\ \&\ N^{n_2}(x=b)\ \&\cdots]\text{'} \in \mu \in \Omega$$

and if $\lambda \in \Omega$ is an alternative to μ, then

(**) $$\text{‘}(Ex)\left[N^{n_1-1}(x=b) \mathbin{\&} N^{n_2-1}(x=b) \mathbin{\&} \cdots\right]\text{’} \in \lambda$$

for $n_i > 1$. (If $n_i = 0$ or $n_i = 1$ for some i, the corresponding conjunct is omitted from (**).)

The need of the other addition to our conditions is brought out by the observation that so far our conditions do not e.g. make the following implication valid:

$$(Ex)\,K_a(b=c \mathbin{\&} K_b p) \supset (Ex)\,K_a(b=c \mathbin{\&} K_c p).$$

The example shows that in the presence of identities their effect on the sentences that express our uniqueness presuppositions (i.e. on (22) and (24)) will have to be taken into account. This does not happen automatically, and these identities have to be taken into account because alternatives to a given world depend merely on the objective identity of its different inhabitants, not on how they are referred to. The best way of doing so seems to me to carry out suitable preliminary simplifications before applying our conditions. I shall not try to formulate the details of these preliminary simplifications here, since the main idea (and its application to many particular cases) is obvious enough. It is important to be aware of the need of further work here, however.

XV. OUR SEMANTICS IS REFERENTIAL

Perhaps the most remarkable feature of this semantics is that one does not quantify over arbitrary functions (or partial functions) that pick out a member from the domain of each possible world (or from some such domains). What one quantifies over is the totality of those functions that pick out *the same* individual from the domains of the different possible worlds (or from some of them). Arbitrary functions of the former kind are essentially what many philosophers call individual concepts, while the latter, narrower set of ‘individuating functions’ essentially represents the totality of the well-defined individuals we can speak of. Thus the ontology of our semantics, indicated by the ranges of the quantifiers we need, is essentially an ontology of ordinary individuals. The main reason why we have to conceive of our individuals as functions is the obvious fact of our conceptual life that our individuals are not determined for one possible course of events or for one particular moment of time only, but can appear in different roles ('embodiments', 'manifestations', or whatever word you want to use) under several of them. For this reason, to speak of an individual is to speak of its different 'embodiments' in different 'possible worlds', which in turn is but to speak of the function which serves to identify these 'embodiments' as manifestations of *one and the same* individual. Thus our semantics is in a very precise sense referential. The crucial entities we need are precisely the ordinary objects to which singular terms refer, and no quantification over 'meanings' is needed. (The *only* novelty is that these objects are considered as potential members of more than one course of events or state of affairs.) We have thus reached an essentially referential semantics.[20] This is all the more remarkable, it seems to me, in view of the fact that the problems (especially the problem of quantifying into modal contexts) which led us to our present treatment are precisely the problems for the treatment of which non-referential notions such as individual concepts, '*Sinne*', etc. were initially introduced. In a fairly strong sense, we have thus eliminated the need of such notions. (Admittedly, some further questions may still persist, for instance questions pertaining to the nature of individuating functions actually used in our ordinary discourse. I shall by-pass them here, although they certainly need further attention.)

In my semi-semantical treatment, this referential character of our theory is signalled by the fact that in $(C.U_m)$ and $(C.E_m)$ the singular terms of whose existence or non-existence we had to speak were precisely those satisfying the uniqueness conditions (24), i.e. the singular terms which specify a real (unique, well-defined) individual. I have already indicated why we can speak in conditions like $(C.U_m)$ and $(C.E_m)$ of the existence of suitable singular terms instead of speaking of the existence of the individuating functions (= 'real' individuals defined for several different possible worlds), which of course is what in the last analysis is involved in the semantics of modality. I have simply assumed that there is a term correlated with each such individuating function and doing the same job of picking out the different incarnations of the same individual.

There may also be some differences, however, between our treatment and the more explicitly semantical discussions of modality that have actually been given. Certain observations in any case seem easier to make within my framework than in some others.

XVI. FURTHER REMARKS. THE RELATIVITY OF THE NOTION OF INDIVIDUAL

It is in my opinion a remarkable fact that essentially the *only* changes in the deductive relationships among one's formulas necessitated by the new semantics are the ones we have carried out by replacing (C.U) and (C.E) by ($C.U_m$) and ($C.E_m$), respectively. Essentially the *only* change needed is therefore to make formulas of the form (22) (or (24)) logically independent of simpler formulas (e.g. of their own substitution-instances). This insight, it seems to me, emerges more readily from my treatment than from some of the competing ones.

Another suggestion which can be elicited from our discussion is that one's notion of an individual is, in a certain sense, relative to the context of discussion. This is brought out by the fact that the formulas (22) and (24) which serve to guarantee that the singular term '*b*' behaves like a name of a genuine individual are relative to a modal profile. In this respect, I find some of the recent semantical treatments of modality far too absolutistic. Normally, we are not interested in the very long 'trans world heir lines' that pick out the same individual from all sorts of possible worlds. Very often, the only things we are interested in are fairly short bits of these lines, and the only quantification that we really need in such circumstances is quantification over these bits. This is the ultimate reason why in the extreme case of quantification in non-modal logic we do not have to worry about cross-world identifications (i.e. about the roles that actually existing individuals may play in other possible worlds) at all. What precisely the uniqueness requirements are that we have to take into account are formulated more readily in my semi-semantical approach than in ordinary semantical theories.

As was already indicated, an extreme form of this relativity of our notion of an individual is in effect the parallelism of existential presuppositions and uniqueness presuppositions. Only when the relativity I am pleading for is acknowledged can we appreciate the important connection which is signalled by the title of my paper and which is expressed by the slogan that modal logics are in the last analysis but so many 'free' logics.

It can also be seen that some of the assumptions we have made are in effect assumptions concerning the behavior of the lines of cross-identification ('world lines' formed by the different embodiments of one and the same individual in different worlds). For instance, (C.ind$=_0$) says in effect that such a trans world heir line never branches (splits) when you move from a possible world to its alternatives, however distant. This assumption is not made in all current semantical systems of modal logic, although it has been vigorously defended by some philosophers of logic, notably by Sleigh.[21] Here we can again see what the semantical counterparts of our assumptions are.

We can also see that they will at least partly justify my earlier statement that Quine has not put his finger quite on the right spot in emphasizing the role of the substitutivity of identity. Interpreted in the way we have done (as an assumption of the parity of identical *individuals*), it is a condition that insures that our concept of an individual behaves in certain desirable nice ways (in that individuals do not 'branch'), rather than an indispensable condition that our concept of an individual must in any case satisfy. (The really essential requirements are those codified by (C.U_m) and (C.E_m).) It seems to me that most of our modal notions (e.g. epistemic modalities) satisfy (C.ind$=$), i.e. behave in these nice ways, but I do not see anything really unique in this particular mode of well-behavedness. It does not seem much more desirable than the converse mode of smooth behavior, which can be described as the impossibility of merging when one goes from a possible world to its alternatives. Yet for most modalities as they are actually used merging seems to me impossible to rule out. (Suppose, for instance, that you have a correct belief as to who or what *a* is, and a similarly correct belief as to who or what *b* is. If the two are in fact different, does it necessarily follow that you must *believe* that they are different? I do not see that this follows at all.)

It must be admitted, however, that there is one fairly strong general reason for ruling out branching in the way Quine has in effect advocated (if I have interpreted him correctly). Earlier, I mentioned the extremely interesting and useful distinction between what can be said of the several references of a term in the different possible worlds we are considering (e.g. of the next president of the United States, 'whoever he is or may be') and between what is said of the individual who in fact (in the actual world) is referred to by a term (e.g. of the man who in fact will be the next president). If we allow branching, the last-mentioned individual is of course not uniquely determined. Then the whole distinction becomes largely inapplicable.

The fact that we can easily understand the distinction and can use it to explicate successfully ambiguities which we feel really are there in our ordinary usage suggests very strongly that the kind of branching we have been discussing is tacitly ruled out in our conceptual system – at least in most circumstances. Hence Quine seems to have after all good reasons for his position, at least in so far as our actual conceptual system is concerned. The naturalness and philosophical interest of the distinction can perhaps be made more obvious by pointing out that to all practical purposes it amounts to the old distinction between modalities *de dicto* and modalities *de re* which was one of the most interesting and useful conceptual tools of the scholastic philosophers.

A closer look at the conditions (C.ind$=_0$) and (C.ind$=$E) in terms of the 'world lines' that our individuating functions define will perhaps enable us to see more clearly what these conditions amount to and what their justification may be. In (C.ind$=_0$) we are considering two 'world lines' both of which are defined for the same selection of possible worlds and which intersect in the actual world. The condition says that these world lines coincide in all the possible worlds in which they were assumed to be defined.

In (C.ind$=$E) we are dealing with two world lines which are defined on partly different classes of possible worlds (which both include the actual one) and which again intersect in the actual world. The condition says that these two world lines can be combined into the world line of one and the same individual.

In both cases, the required behavior of 'world lines' is such as we clearly would like our notion of individual to exhibit. In this sense, there is a great deal to be said for them. Whether it is realistic to assume that we can actually have as nicely defined a notion of individual in the presence of each important modal notion is a question which cannot be adequately discussed here. Nor can it be disposed of by bland assertions to the effect that without those conditions one cannot 'understand' or 'make sense of' quantification into modal contexts. Our semi-semantical treatment already gives us hints as to what the sense would be. Nevertheless, there are good reasons – especially those derived from the successful applications of the *de dicto* – *de re* distinctions – to suggest that in most cases the assumptions are applicable to most of our own modal concepts.

XVII. WORLD LINES CANNOT ALWAYS BE CONTINUED

This problem is connected with another aspect of our approach which at first sight might seem disconcerting. It may seem strange that certain sentences of the following form are *not* always logically true:

(49) $(Ex)p \supset (Ex)(p \,\&\, q)$

where q 'seems' to follow easily from p in the sense that the statement $p(a/x) \,\&\, \sim q(a/x)$ is inconsistent when a is a new individual constant. The technical reason is clear enough. Part of a counter-example to (49) would look like this:

(50) $(Ex)p \in \mu \in \Omega$

(51) $(Ux)(\sim p \vee q) \in \mu$

(52) $p(a/x) \in \mu$

(53) $`Q_p(a)' \in \mu.$

Here '$Q_p(x)$' is the uniqueness premise which goes together with the modal profile of p with respect to 'x'. Here (52)–(53) of course follow from (50) by (C.E$_q$). The only further conclusion that can be drawn here is

(54) $[\sim Q_{p\&q}(a) \vee \sim q(a/x)] \in \mu$

(from (51) by (C.U$_q$)). If we could rule out the possibility that

(55) $`\sim Q_{p\&q}(a)' \in \mu$

we could produce an inconsistency, for then we would have

(56) $\sim q(a/x) \in \mu$

by (C. $\vee$). However, there is nothing to rule out (55), provided only that the uniqueness requirement expressed by '$Q_{p\&q}(x)$' is stronger than that expressed by '$Q_p(x)$'. (The only way of ruling out (55) seems to make (53) imply that '$Q_{p\&q}(a)$' $\in \mu$.) Hence a counter-example cannot be ruled out, and (49) need not be logically true.

Does this go to show that our treatment of modality is unnatural? In my opinion it does nothing of the sort, however surprising and disconcerting this phenomenon might first appear. Rather, it gives us a chance of characterizing interesting differences between different kinds of modal notions.

The semantical situation that goes together with the failure of the logical truth of (49) is not hard to fathom. The relativization of uniqueness presuppositions means that normally we are not quantifying over the whole long trans world heir lines that constitute all that can be said of our individuals in objective terms, but rather quantifying over assorted bits and pieces of such heir lines.

If each such bit of an heir line could always be extended arbitrarily far to further possible worlds, this would not make any difference. But assuming this would mean assuming further interconnections between the different uniqueness conditions. These assumptions amount to assuming the validity (logical truth) of sentences of the following form:

(57) $\text{`}(Ux)\,[Q_1(x) \supset (Ey)\,(x = y \;\&\; Q_2(y))]\text{'}$

for some (or maybe all) the different kinds of uniqueness premises 'Q_1', 'Q_2'. These very same assumptions are, as our example (50)–(56) above suggests, just what is needed to show such implications as (49) to be logically true.

Such assumptions as (57) might look very tempting. This temptation is probably due to the fact that statements of form (57) are in fact logically true *for logical necessity* and perhaps also for physical (natural) necessity. However, there is no reason to assume the validity of (57) for most of the other modal notions (in the wide sense of the word), including propositional attitudes. In fact, an individual might e.g. be perfectly well defined as far as the belief-worlds of some specified person (say the person referred to by 'a') are concerned (and hence give rise to nice trans world heir lines connecting these worlds), and yet fail to be uniquely determined as far as somebody else's beliefs are concerned – which means that the heir lines in question cannot be extended to *his* 'belief worlds' (worlds compatible with everything he believes). The same holds obviously for many other propositional attitudes.

An example will hopefully convince the reader of the relevance of what I just said. An instance of (57) might be

(58) $\text{`}(Ux)\,[(Ey)\,B_a(y = x) \supset (Ez)\,(z = x \;\&\; (Ey)\,B_a\,B_b(y = z))]\text{'}$

which would say (roughly) that whenever the bearer of 'a' has an opinion concerning the identity of an individual, he believes that the person referred to by '*b*' also has such an opinion. This, of course, is clearly false in most cases.

The presence of such further assumptions as the transitivity and reflexivity (of the alternativeness relation which goes together with a modal notion) greatly reduces the number of possible assumptions of form (57) to be accepted or rejected. But even then one can see that these assumptions are generally unacceptable. For instance, for a discussion of a particular man's knowledge (let him be referred to by '*a*'), one of the few relevant assumptions of form (55) would be expressed by

(59) $\text{`}(Ux)\,[(Ey)\,(y = x) \supset (Ez)\,(z = x \;\&\; (Ey)\,K_a(y = z))]\text{'}\,.$

What this says is that the man referred to by '*a*' knows of each actually existing individual who or what it is. This is obviously false in all interesting applications.

Thus the failure of statements of form (49) to be logically true is not disconcerting at all. Instead, it points to an interesting general conceptual fact. Such statements as (49) and (57) fail to be logically true because what counts as an individual varies from one man and from one attitude to another, and is not determined by the set of actually existing individuals. Thus the failure of an instance of (49) or (57) for a given notion seems to be an indication of the intentional (psychological) character of the notion. In contrast to such intentional notions as belief, non-intentional modalities like logical necessity seem to make them logically true.

If assumptions like (57) are combined with (C.ind=) and (C.ind=E), all uniqueness premises will coincide with each other and with the existence premise '$(Ex)(x=b)$'. The situation then becomes rather trivial. This fact may perhaps be used as an argument against the interest of logical modalities as an object of semantical study compared e.g. with propositional attitudes. More generally, in this way we can see, not just the technical possibility of relativizing one's uniqueness assumptions, but some of the insights gained by so doing – and the necessity of such a course in the case of propositional attitudes.

XVIII. CONTRA LOGICAL MODALITIES

This line of thought can in fact be turned into a more serious criticism of logical modalities than a comment on their relatively trivial logical behavior. We have not yet said anything about the ways in which the identifying functions can be defined which enable us to speak of the *same* individual appearing in several different possible worlds. Nor can anything like a satisfactory discussion be given within the confines of one essay. However, it is clear what sorts of criteria are used here: they turn on the similarities between different possible worlds and on regularities obtaining in each of the possible worlds we have to consider (for instance, on the continuity of our individuals with respect to space and time). If this is not immediately clear to a reader, we can invite him to consider what it is that makes it possible for him to speak of more or less the same set of individuals all the time when discussing what possible courses of events might materialize between today and next week, as far as his beliefs are concerned. If he did not believe in the spatial and temporal continuity of persons, chairs and molecules, he might have some difficulty in justifying his talk of the *same* individuals independent of the particular course of events he happens to be considering.

If this is the case, relativization of world lines is a dire necessity. For if we want to extend them indefinitely, we might run into possible worlds that simply are so irregular that our customary methods of cross-identifying individuals (=telling whether the inhabitants of different possible worlds are or are not the same) may simply fail.

In the case of most applications of propositional attitudes, this is avoided because the possible worlds that are in fact compatible with people's propositional attitudes are fairly regular and pretty similar to each other. However, it seems to me that even here the applicability of our semantical concepts depends on assumptions concerning the degree of realism in people's propositional attitudes. But in the case of logical modalities (logical and analytical possibility and necessity) the different worlds we (so to speak *per definitionem*) have to consider can be so irregular and dissimilar that all the methods of cross-identification that are used in our native conceptual system are bound to fail. If so, we cannot quantify into contexts governed by words for such logical modalities, for such quantification depends essentially on criteria of cross-identification (individuating functions, world lines).

If so, Quine turns out to have been right in his suspicion of quantified modal logic in the narrow sense of the word as quantification theory plus *logical* modalities. However, there do not seem to be any objections to a theory of propositional attitudes *cum* quantification. In fact, it is only by developing a satisfactory semantical theory for languages which embody both these elements that the deep true reasons for rejecting a quantified logic of logical modalities finally begin to emerge.

REFERENCES

[1] The Irvine Colloquium in May 1968.

[2] For further remarks on this point, see my paper, 'Logic and Philosophy' in *Contemporary Philosophy – La Philosophie Contemporaine*, vol. I (ed. by R. Klibansky), Florence 1968.

[3] 'Language-Games for Quantifiers', *American Philosophical Quarterly, Monograph Series*, no. 2 (1968): Studies in Logical Theory, pp. 46–72.

[4] On this subject, see my paper 'On the Logic of Existence and Necessity I: Existence', *The Monist* **50** (1966) 55–76, reprinted in the present volume as 'Existential Presuppositions and Their Elimination'. The present paper includes much of the material which I intended to include in the second part of that earlier paper.

[5] There will be some overlap with my discussion note 'Individuals, Possible Worlds, and Epistemic Logic', *Noûs* **1** (1967) 33–62.

[6] See e.g. 'Knowledge, Identity, and Existence', *Theoria* **33** (1967) 1–27; 'Interpretation of Quantifiers' in *Logic, Methodology, and Philosophy of Science III, Proceedings of the 1967 International Congress* (ed. by B. van Rootselaar and J. F. Staal), Amsterdam 1968, pp. 435–444; also 'Quine on Modality', *Synthese* **19** (1968–69) 147–157.

[7] In order to prove this, it suffices to show that whenever Ω is a model system which satisfies the earlier conditions, we can adjoin new formulas to its members so as to obtain a new model system Ω' which in addition to the earlier conditions also satisfies (C.N=). (Then the same sets of formulas will be satisfiable in either case.) This can be accomplished as follows: whenever $\mu \in \Omega$, adjoin to μ all formulas p such that for some finite sequence of formulas $p_0 = p, p_1, p_2, \ldots, p_k$ and some suitable singular terms 'a_1', 'b_1', 'a_2', 'b_2',..., 'a_k', 'b_k' (not necessarily different), '$N^{n_i}(a_i = b_i)$' $\in \mu$ or '$N^{n_i}(a_i = b_i)$' $= p_j$ for $j < i (i = 1, \ldots k)$ and p_i and p_{i-1} are like except that a_i and b_i have been exchanged at some place or places where they occur in the scope of precisely n_i modal operators. That Ω' so constructed satisfies (C.N=) is immediately obvious. That it satisfies the other conditions can be proved by induction on the number of symbols '&', '$\vee$', 'E', 'U', 'N', 'M'.

[8] See e.g. my paper, 'Modality and Quantification', *Theoria* **27** (1961) 119–128.

[9] Admittedly Quine also frequently mentions the failure of existential generalization as an indication of trouble in quantified modal logic. The impression he leaves, however, is that this is just another symptom of one and the same illness. We shall soon see that the question of the validity of the substitutivity of identity is largely independent of those changes in the quantifier conditions (C.E) and (C.U) which determine the fate of existential generalization.

See e.g. W. V. Quine, *From a Logical Point of View*, 2nd ed., Cambridge, Mass., 1961, pp. 139–159; *The Ways of Paradox*, New York 1966, pp. 156–182; Dagfinn Føllesdal, papers referred to in note 6 above.

[10] Alternatively, we may express the last part of this condition as follows: ... and if '$Q^{n_1, n_2, \ldots}(b)$' $\in \mu$, then $p(b/x) \in \mu$.

[11] Special cases of this argument were given in Jaakko Hintikka, 'On the Logic of Existence and Necessity' (note 4 above) and 'Individuals, Possible Worlds, and Epis-

temic Logic' (note 5 above). In the latter, the generalization presented here was also anticipated.

[12] Again, (C.U1) may be formulated as follows: ... and if '$(Ex)(N^{n_1}(x=b)$ & N^{n_2} $(x=b)$ &...)' $\in \mu$, then $p(b/x) \in \mu$. Here instead of '$(Ex)(N^{n_1}(x=b)$ &$N^{n_2}(x=b)$ &...)' we may have any formula obtained from it by the following operations: changing the order of conjunction members and/or identities; replacing the bound variable everywhere by another one.

[13] Here we see especially sharply the difference between questions pertaining to the substitutivity of identity and questions pertaining to existential generalization. In the former, the question is whether two singular terms pick out the same individual in *each* possible world in a certain class of possible worlds (considered alone without regard to the others). In the latter, we are asking whether a given singular term picks out one and the same individual in *all* possible worlds of a certain kind (when they are compared with each other).

[14] See e.g. my paper 'Modality and Quantification' (note 8 above).

[15] The general validity of (26) presupposes that any actually existing individual also exists in all the alternatives to the actual world. The following model system Ω provides a counter-example to (26): Ω consists of μ and ν, the latter of which is an alternative to the former. Here

$$\mu = \{(Ex)Np, Np(a/x), (Ex)(x=a), p(a/x), M(Ux) \sim p$$
$$\nu = \{(Ux) \sim p, p(a/x)\}.$$

We could not have this counter-example, however, if '$(Ex)(x=a)$' $\in \mu$ entailed '(Ex) $(x=a)$' $\in \nu$, i.e. if we could 'move' an existence assumption concerning a from a possible world to its alternatives.

[16] Thus the elimination of existential presuppositions helps us to dispense with unwanted assumptions concerning the 'transfer' of individuals from a possible world to its alternatives.

[17] This is one of the many places where one is easily misled if one trusts uncritically the superficial suggestions of ordinary language. Surely there are circumstances in which someone knows who is referred to by '*a*' is and also knows who is referred to by '*b*' is while in reality '$a=b$' is true, apparently without thereby *knowing that the references of 'a' and 'b' are identical*, contrary to what (C.ind.$=_0$) requires. However, one has to insist here very strongly that in the two cases of a and b, respectively, precisely the same sense (same criteria) of knowing who must be presupposed. This is not the case, it seems to me, in any of the apparent counter-examples that have been offered.

[18] In this paper, 'Some Problems about Belief' *Synthese* **19** (1968–69) 158–177, especially pp. 168–169, Wilfrid Sellars claims in effect that the validity of (C.ind$=_0$) is ruled out by the interpretation of quantifiers which I propose in my *Knowledge and Belief*, Ithaca, N.Y., 1962. This argument completely misconstrues my intended interpretation, however, for reasons I can only guess at, and hence fails to have any relevance here. Although (C.ind$=_0$) was not mentioned in *Knowledge and Belief*, there is nothing there that rules this condition out for syntactical or for semantical reasons. Nor is there anything in *Knowledge and Belief* that is affected by the adjunction of this new condition.

[19] A few additional comments are presented in my paper, 'Semantics for Propositional Attitudes' in *Philosophical Logic* (ed. by J. W. Davis, D. J. Hockney and W. K. Wilson), D. Reidel Publishing Company, Dordrecht 1969, reprinted in the present volume, pp. 87–111.

[20] Cf 'Semantics for Propositional Attitudes' (note 19 above).

[21] See R. Sleigh, 'On Quantifying into Epistemic Contexts', *Noûs* **1** (1967) 23–32.

IV. CONCEPTUAL ANALYSES

ON THE LOGIC OF PERCEPTION

I. THE LOGIC OF PERCEPTION AS A BRANCH OF MODAL LOGIC

Should the title of this paper prompt you to ask, "What *is* the logic of perception?", there is an answer at hand. I shall argue here that the logic of our perceptual terms is a branch of *modal logic*.[1] In saying this, I mean by 'perceptual terms' both such words as 'sees', 'hears', 'feels', etc., which involve a reference to one particular sense modality, and such words as 'perceives', which are neutral in this respect. By modal logic, I mean not only the logic of the terms 'necessary' and 'possible' but also the logic of all the other terms that can be studied in the same ways as they. Among these terms are most of the words that are usually said to express propositional attitudes, including 'knows', 'believes', 'remembers', 'hopes', 'strives', etc. What is in common to all the modal notions in this extended sense of the term will be partly explained later.[2]

The close relation between theory of perception and modal logic will be argued on two levels. First, I want to outline the basic reasons why the logic of perception is susceptible to the same sort of treatment as other modal logics. Secondly, I want to show, by discussing a number of interrelated problems, that by treating perceptual concepts as modal notions we can shed sharp new light on some of the classical issues in the philosophy of perception. These include the evaluation of the so-called argument from illusion, the status of sense data, and the nature of the objects of (immediate) perception. If I am right, there are interesting connections between these problems and some of the questions we are led to ask when we study the semantics of modal logic in general. Personally, I came to appreciate for the first time the theoretical interest of some of the traditional philosophical problems concerning perception when I realized that they are in effect identical with or at least very closely related to the difficulties logicians and philosophers of logic have recently encountered in trying to understand the interplay between modal notions and the basic logical concepts of identity and existence (of quantification). Problems drawn from the field of perception can even serve to illustrate and to elucidate these general logical and semantical difficulties.

Some of the formal similarities between perceptual terms and other modal terms are obvious enough, so obvious in fact that I find it surprising and perplexing that they should have been registered, to the best of my knowledge, only once in the relevant philosophical literature. In her Howison Lecture, 'The Intentionality of Perception: A Grammatical Feature', Miss G. E. M. Anscombe points out a number of similarities between perceptual concepts and concepts she calls intentional.[3] Most of these similarities hold between perceptual notions and modal notions in general. An interesting example is the distinction which Miss Anscombe calls the difference between material and intentional objects (of an activity). A man aims his rifle and fires at a dark patch against the foliage which he takes to be a stag (intentional object). Unknown to him, and most unfortunately, the dark patch was his father (material object), a fact which causes a tragedy. This tragedy brings out the importance of the difference between the intentional object aimed at and the material object aimed at. The same distinction clearly applies to perception. Miss Anscombe connects it with Austin's contrast between "Today I saw a man shaved in Oxford", and "Today I saw a man born in Jerusalem", both uttered in Oxford.

The description and perhaps also the formalization of such features of the logic of perception is an interesting and worthwhile enterprise. I do not believe, however, that it can be really successful until we have deeper insights into the way modal notions function and into the reasons why perceptual terms are in this respect like other modal terms.

Modal notions, including propositional attitudes, can be classified according to the degree (kind) of success they presuppose. For instance,

a knows that p

cannot be true unless it is the case that p, while, e.g.,

a believes that p
a hopes that p

can both be true even though it is not true that p.

The question of the status of perceptual verbs vis-à-vis this distinction is not quite clear, and requires a word of warning. As Gilbert Ryle points out in *The Concept of Mind*,[4] the words 'perceive' and 'perception' as well as 'see', 'hear', etc., are normally used "to record observational success". Most of the discussion in this paper is neutral with respect to this distinction, however. A success presupposition is to be read into my use of perceptual terms only when an explicit statement is made to this effect. (In fact, I want to suggest that the interest of the success presupposition and of the correlative possibility of perceptual mistakes is often overrated.) Thus it might be more natural to use locutions like 'it appears to *a* that *p*' rather than '*a* perceives that *p*'. This would also serve to bring out more clearly what I want to focus on in this paper: the problems connected with one's description of one's immediate perceptual experience (no matter whether it is veridical, unwittingly misleading, or an acknowledged illusion). However, for simplicity, I shall normally employ the shorter locution 'perceives that'.[5]

II. MODAL LOGIC AS TURNING ON THE NOTION OF 'POSSIBLE WORLD'

It seems to me that the best way of achieving conceptual clarity in modal logic is to view all the use of modal notions as involving a reference, usually of course only a tacit one, to more than one possible state of affairs or course of events (in short, to more than one 'possible world'). In this way, most of the conceptual problems that philosophers of logic have run into in this area become manageable.[6] There is basically nothing unusual or strange in the relation of our terms to their references in modal contexts, I suggest. What seems to cause problems is merely the fact that in such contexts a term may have a reference in more than one 'possible world', i.e., that we have to consider our individuals as (potential) members of more than one state of affairs or course of events. Modal contexts thus do not exhibit any *failure of referentiality*, but only *referential multiplicity*. From this point of view, the logic of perception can also be elucidated.

This procedure will be made clearer by my subsequent remarks. There are a few possible misunderstandings, however, against which I first must guard myself. First of all, there is nothing mysterious about what I have called 'possible worlds'. Following Richard Jeffrey,[7] we might call a 'complete novel' a set of sentences in some given language which is consistent but which cannot be enlarged without making it inconsistent. A possible world is in effect what such a complete novel describes. Actually, many useful purposes are served by descriptions of possible worlds that are less than complete, as long as these partial descriptions are large enough to show that the world they purport to describe is really possible. Such partial descriptions of possible worlds I have called 'model sets', and I have discussed them in some detail elsewhere.[8]

Second, it is important to make a distinction between two different cases here. Sometimes (e.g., in the case of the concepts of possibility and necessity) what I have called 'possible worlds' are normally different possible *courses of events*. Sometimes (e.g., in the case of perceptual concepts) the relevant 'possible worlds' are normally different possible *states of affairs* at the particular moment of time we are talking about. This distinction does not affect what I shall say in the sequel, however.

Third, I am of course not suggesting that the ordinary people who daily use such ordinary words as 'sees' or 'hears' or 'possible' or 'necessarily' are ever interested in anything as fancy as 'possible worlds' or 'possible states of affairs'. Surely what they are interested in is just the unique world of ours that happens to be actualized. The point is, rather, that many of the things we all say daily about this actual world of ours can be explicated by a logician in terms of his 'possible worlds'. A logician might say that we often succeed in saying something about the actual world only by locating it, as it were, on the map of all the different possible worlds.

This is by no means restricted to modal concepts. If I understand a prediction, I know which future courses of events are such that the prediction can be said to have been successful under them, and which courses of events are such that the prediction will have to be said to have failed under them. If I know that the prediction is true, I know that the course which events will actually take ('the actual world') is of the first kind and not of the second. In general, understanding a sentence is being able to divide all possible worlds into two classes: those in which the sentence would be true, and those in which it would be false. For certain technical purposes, the sentence (or the proposition it expresses) might even be identified with the former set of possible worlds. The reason why this gambit succeeds so often should be clear: One understands what a sentence says in so far as one knows what to expect of the world in case the sentence is true.

What is peculiar about modal concepts is only the fact that in order to spell out their logic we have to consider several possible worlds in their relation to each other, and not just one possible world at a time, as we can do in explaining the semantics of ordinary, non-modal logic. The intuitive reason for this difference is that in order to explain what it means for a non-modal statement to be true in a possible world it suffices to consider that world only, whereas the truth conditions of a modal statement cannot be spelled out without considering possible worlds other than the one in which it is supposed to be true. For instance, 'possibly *p*' can be true in the actual world only if *p* were true in some (suitable) possible worlds. In other respects, almost the same things could be said of modal sentences as were said above of non-modal sentences. I know what someone believes if and only if I can tell those possible worlds which are compatible with everything he believes from those which are incompatible with his beliefs. I know what somebody sees at a given moment of time in so far as I can distinguish between states of affairs (at that moment of time) which are compatible with what he sees and states of affairs which are incompatible with his visual perceptions at the time. One can virtually paraphrase all attributions of such 'propositional attitudes' as knowledge, belief, wish, hope, perception, etc. to someone in terms of possible worlds compatible with his attitudes at a given moment of time. For instance, we may tentatively put:

(1) *a* believes that *p* = in all possible worlds compatible with what *a* believes it is the case that *p*;

(2) *a* does not believe that *p* (understood in the sense 'it is not the case that *a* believes that *p*') = there is a possible world compatible with what *a* believes in which not-*p* would be true;

(3) *a* perceives that *p* = in all possible states of affairs compatible with what *a* perceives it is the case that *p*;

(4) *a* does not perceive that *p* (understood in the sense 'it is not the case that *a* perceives that *p*') = there is a possible state of affairs compatible with everything *a* perceives in which not-*p* is true.

If it is objected that these equations are but so many tautologies, my answer will be that they are intended to be just that. Their right-hand sides were intended to be merely paraphrases that do not add anything to the expressions on the left-hand side but are in a form somewhat more conducive to conceptual clarity than the original formulation.[9]

One may object that such paraphrases as (1) have the paradoxical consequence that anybody who believes (knows, remembers, etc.) that *p* believes (knows, remembers, etc.) all the logical consequences of *p*. This objection is easily parried, however, by understanding '*q* is compatible with what *a* believes' as meaning '*q* is not an analytic consequence of *p*', where *p* is a formulation of what *a* believes and where the notion of analytic consequence is to be understood in one of the senses explained in my paper 'Are Logical Truths Tautologies?'[10]

Although such paraphrases as (1)–(4) are thus somewhat crude, they bring out several relevant features of the conceptual situation. From them we can see that whenever we are discussing, say, the beliefs of a given person, the possible worlds we have to consider are the possible worlds compatible with his beliefs. In general, whenever we ascribe or deny a given propositional attitude to a person (with respect to any propositions whatsoever), the possible worlds we have to consider are those compatible with the relevant propositional attitude of his.[11]

From these paraphrases we can also see in sharper detail what forces us to consider several possible worlds in one and the same 'logical specious present'. It is the possibility of disclaiming a propositional attitude, as in (2) and (4) that necessitates this. If it is said that someone, say *a*, does not believe that *q* and that he also does not believe that *r*, this will be tantamount to saying that there is a possible world compatible with everything *a* believes in which *q* would be false and that there is also a similar possible world in which *r* would be false. There is absolutely no reason for supposing that these two possible worlds are identical, and very often it is on the contrary obvious that they are not. For instance, it is perfectly possible that $r = \text{not-}q$; this merely means that *a* does not have the opinion that *q* nor the contrary opinion that not-*q*. Then requiring that the two possible worlds be identical means requiring that there be a possible world compatible with everything *a* believes in which both *q* and not-*q* would be true, which violates the principle of non-contradictoriness. Hence we often have to consider more than one possible world compatible with someone's propositional attitudes. Only in the case of an omniscient *a* can we restrict our attention to one world only.

We can see perhaps how intricately propositional attitudes are involved conceptually with the notion of a possible world by asking what it means for someone's propositional attitude to be more extensive than another person's similar attitude. When does *a* know (believe, wish, perceive) more than *b*? The only reasonable general answer seems to be that *a* knows more than *b* if and only if the class of possible worlds compatible with what he knows is smaller than the class of possible worlds compatible with what *b* knows; and similarly for the other propositional attitudes. This is not a full answer by any means, for it does not tell us yet how the different possible worlds are to be separated from each other and how they are to be weighted in relation to each other. It suffices to show, nevertheless, how important the notion of a possible world is for our understanding of the logic of propositional attitudes.

III. QUANTIFICATION AND IDENTITY IN MODAL CONTEXTS

This notion is especially useful in clearing the conceptual muddles that have beset recent attempts to understand the interplay of propositional attitudes and other modal notions with such basic logical concepts as the quantifiers (*Ex*) ('there is at least one individual, call it *x*, such that') and (*Ux*) ('of each individual, call it *x*, it is true that') and the concept of identity (of individuals) '='. The problems that arise in this area are epitomized by the breakdown of the modes of inference known as 'substitutivity of identity' and 'existential generalization'.[12] The former says, somewhat roughly expressed, that whenever an identity of the form '$a=b$' is true, the terms '*a*' and '*b*' are interchangeable everywhere *salva veritate*. The latter says that if a statement containing a free singular term, say '$F(a)$', is true, then so is the result '$(Ex)F(x)$' of replacing this free singular term by a variable bound to an existential quantifier. The striking thing about these two modes of inference is that they seem to be obviously and undoubtedly valid in so far as the only task of our singular terms is merely to refer to the individuals we are talking about. For, if this is the case, how could these two modes of inference go wrong? If two individuals are identical, must not exactly the same things be true of them both? If something is true of the particular individual specified by the term '*a*', must not this something be true of some individual or other?

Yet these two modes of inference break down in modal contexts. George IV knew that Walter Scott was Walter Scott. Furthermore, Walter Scott was the author of *Waverley*. Nevertheless, the good king did not know that he was, although this would follow by the substitutivity of identity.[13] I may hope that the next governor of California is a Democrat, but it does not follow from this that there is some particular Democrat whom I hope to see elected, contrary to what existential generalization would suggest that I do.

Different philosophers have reacted to this predicament differently. Some have taken the breakdown of these inferences, together with certain further observations, to show that the values of bound variables in modal contexts are something different from the ordinary 'extensional' entities they usually range over. They have declared that if all these variables did was simply to range over such entities, existential generalization would have to be applicable in these contexts. Others, wary of any unusual 'intensional entities', have wanted to explain away those uses of modal concepts which cause the breakdown of the problematic inferences. A frequent device of these philosophers has been the postulation of different senses of propositional attitudes.[14] Some of these senses are allegedly free of the difficulties we have been discussing. A few philosophers have gone so far as to suggest that these 'extensional' or 'transparent' senses of modal concepts are all that there is to the use of quantification in modal contexts. From the point of view here adopted, this amounts to an attempt to deal with modal concepts as if they could be reduced to concepts that involve a reference to the actual world only, and not to any alternatives to it.

Not many words are needed, however, to restore our confidence in logic without postulating either intensional entities or irreducibly different senses of propositional attitudes. It suffices to recall that each modal notion involves a tacit reference to more than one possible world. The actual truth of the identity '$a=b$' means that the terms '*a*' and '*b*' refer to the same individual in the actual world. From this it follows that they are interchangeable in so far as we are speaking of the actual world only, that is to say, in so far as they occur outside the scope of all modal terms. But since modal terms introduce more than one possible world and since there are no general reasons why two terms (like '*a*' and '*b*') that actually refer to one and the same individual should do so in other possible worlds, there is not the slightest excuse to think that they are interchangeable in modal contexts.

This is just what is illustrated by our example. The reason why 'Walter Scott' and 'the author of *Waverley*' are not interchangeable, although they refer to the same person, was that the good king did not know that their references are identical, i.e., that in one of the possible worlds compatible with everything George IV knew, Walter Scott is not the author of *Waverley*.

Similarly, the reason we cannot always generalize existentially with respect to a free singular term in a true sentence like '$F(a)$' is that the term in question (our '*a*') may refer to different individuals in the possible worlds that are brought to play by the modal notions that occur in '$F(a)$'. If it refers to different individuals in this way, there is no one individual of whom (or which) we are speaking when we say that $F(a)$, and therefore there is no foothold for maintaining that there is some *individual* (say x) who is such that $F(x)$. This is again exactly what happened in my example: Under the different courses of events compatible with my present hopes, different men will be elected, that is to say, the term 'the next governor of California' refers to different individuals in the different 'possible worlds' compatible with what I hope to happen. Hence it is not amenable to existential generalization.

Thus the breakdown of existential generalization and of the substitutivity of identity in modal contexts is not a symptom that our free singular terms refer to entities different in kind from their normal references. Rather, the breakdown is a direct consequence of the fact that in modal contexts we have to consider our individuals as members of more than one state of affairs or course of events.

We can also see at once how the problematic inferences are to be restored by means of supplementary premises. In order for the terms '*a*' and '*b*' to be interchangeable, they have to refer to the same individual not just in the actual world but also in all the other 'worlds' we are considering. For instance, if we are speaking of what *d* believes, these additional worlds are those compatible with everything he believes. The substitutivity of '*a*' and '*b*' thus requires more than the truth of '$a=b$'; it also requires the truth of '*d* believes that $(a=b)$'. Other modalities behave likewise.

To restore existential generalization, we have to assume that the term with respect to which we are generalizing, say 'a', refers to the same individual in all the different 'worlds' we are considering. They are the actual world plus whatever possible worlds are compatible with the relevant propositional attitudes of the person we are talking about (these do not always include the actual world). The requirement that this should be the case can again be expressed by an explicit premise. In the case of belief it will be

(5) (Ex) (d believes that $(a = x)$ and $(a = x)$),

where d is the man we are talking about. The first conjunct in (5) makes sure that the term 'a' refers to one and the same individual in all the possible worlds compatible with what d believes, and the second conjunct guarantees that this uniqueness of the reference of 'a' extends to the actual world. Analogously, existential generalization is reinstated in perceptual contexts by premises of the form

(6) (Ex) (d perceives that $(a = x)$ and $(a = x)$),

where instead of the word 'perceives' we could also have one of the more specific words like 'sees'. Occasionally (for instance, when the veracity of d's perceptions is not at issue, directly or indirectly) we can use instead of (6) the simpler premise

(7) (Ex) (d perceives that $a = x$).

This premise can be used instead of (6) also if it is required that perceptual terms have a success grammar, that is to say, if it is required that one can perceive only what is in fact the case.

Expressions of form (7) or (6) are extremely interesting in the logic of perception, as their analogues are in other branches of modal logic. Since the effect of (6) is to guarantee that the free singular term 'a' refers to one and the same individual in all the possible states of affairs we have to consider, its import may be expressed somewhat inaccurately but nevertheless strikingly by saying that what (6) says is that a is a genuine (unique) individual in so far as d's perceptions are concerned. This way of bringing out the import of expressions like (6) is perhaps even more natural in the case of some of the other modalities. As far as my hopes are concerned, the next governor of California is a (unique) *individual* if and only if there is some one politician who I hope will be elected. As far as your knowledge is concerned, the prime minister of Norway is a unique individual if and only if there is someone who you know is the Norwegian prime minister, in short, in so far as you know who the prime minister of Norway is.

Statements of the form (5) through (7) might also be called 'identification statements' (in one possible sense of this expression). There is a sense which one has identified the reference of a term, say '*a*', if and only if one knows which individual '*a*' refers to. Likewise, *d* can be said to have perceptually identified *a* in so far as he perceives which individual *a* is, that is to say, in so far as (7) is true. In the case of belief, the simpler identification statement '(Ex) (d believes that $a=x$)' says that *d* thinks (believes) that he has identified *a* while the fuller identification statement (5) says in addition that he is in fact right in his belief about the identity of *a*.

These observations should make it clear that in our treatment of quantification into modal contexts we are not relying on any unusual sense of quantifiers, e.g., a sense to be defined in terms of the truth of substitution-instances of quantified sentences, as some philosophers have tried to do. For instance, the existential quantifier '(Ex)' is here taken to express precisely the *existence* of a (genuine, i.e., unique) *individual.* It is precisely our insistence on this (normal) sense of quantifiers that necessitates the use of additional premises which serve to guarantee that the free singular terms with respect to which we want to quantify really specify unique individuals capable of serving as values of bound individual variables.[15]

This is also what enables us tc spell out a part of Miss Anscombe's distinction between the intentional and the material objects of perception. When a free singular term occurs within the scope of a perceptual term in a sentence, it specifies an intentional object. However, if this singular term is replaced by a variable bound to an initial quantifier, we obtain a new statement which is no longer about the intentional object, but about the unique individual which as a matter of fact is being perceived, for those are the entities that bound individual variables range over. If it is the case that the relevant value of this variable is a certain individual *b*, then this individual may be said to be the material object of perception. A paradigm of this distinction is the difference between '*d* sees that *d*'s brother is being shaved' and '(Ex) (($x = d$'s brother) and (d sees that x is being shaved))', where the former can only be true if d sees that it is his brother who is being shaved, while the latter may be true even when d does not see that the man whom he sees being shaved is in fact his brother. The distinction at any rate catches some of the things Miss Anscombe apparently wants to say of the difference between intentional and material objects of perception, and of the attitudes philosophers have taken to them. It is readily extended to other examples and to other modalities.

The distinction which is illustrated by our paradigm could be called a distinction between statements about, say, a, 'whoever he is or may be', and statements about the individual (e.g., person or object) who in fact is a. This distinction is closely related to our concept of an individual. Only statements of the second kind can really be said to be *about* definite individuals.

IV. THE ARGUMENT FROM ILLUSION IS ILLUSORY

How, then, do these observations shed new light on the traditional problems concerning perception? What, for instance, do they imply concerning the status of so-called sense-data, which many philosophers have postulated as objects of immediate perception?

What are these sense-data supposed to be and why do we have to assume them in addition to the ordinary physical objects? I suspect that entirely different things have been included by different philosophers among sense-data. There is a line of argument, however, that once was pretty generally taken to show the indispensability of sense-data in the theory of perception. It is generally known as the 'argument from illusion'.[16] It has been put forward in several variants. Disregarding the differences between these variants and the subtleties that prompted these differences, we may say that the argument from illusion consists in inferring the necessity of postulating sense-data from the possibility of illusion or perceptual error. In the case of an erroneous perception, for instance in the case of a perception that shows a red object to be grey, the object of one's (immediate) perception cannot (according to this line of thought) be the physical object in question, precisely because its attributes are different from those of the physical object in question.

Yet our perceptions are *about* something – there *is* something grey that I do sense even in the case of the illusion we are envisaging. Hence we must assume the existence of non-physical objects of immediate perception, at least in the case of an illusion. But since there is no intrinsic difference between illusory and veridical perception, it is argued, their objects have to be similar. In both cases, therefore, the objects of immediate perception must be different from ordinary physical objects. These extraordinary objects of immediate perception are then dubbed sense-data.

Although this sketch of the argument from illusion is so brief as to appear a caricature, it brings out some of the relevant features of this line of thought. Presented in this way it seems to me to be completely devoid of force. The basic mistake, or one of the basic mistakes, lies in the vagueness of contrast between the perceived and the real attributes of an object. On any reasonable view of the matter, be it phenomenalistic or realistic, some distinction has to be made between the experienced (phenomenal) qualities and relations of things and their physical qualities and relations. Usually this distinction is completely disregarded in the argument from illusion. Yet it is absolutely fatal to many forms of the argument, as Thomas Reid already saw. He formulated a special case of the argument from illusion as follows: "The table which we see, seems to diminish as we remove farther from it; but the real table, which exists independently of us, suffers no alteration. It was, therefore, nothing but its image which was presented to the mind." His reply is: "Let us now suppose, for a moment, that it is the real table we see: Must not this real table seem to diminish as we remove farther from it? It is demonstrable that it must. How then, can this apparent diminution be an argument that it is not a real table? When that which must happen to the real table... does happen to the table we see, it is absurd to conclude from this that it is not the real table we see."[17]

Our insight into the nature of perceptual terms as expressing propositional attitudes enables us to point out other mistaken presuppositions in the 'argument'. Underlying it is obviously the idea that perception is to be construed as a simple two-term relation between the perceiver and the perceived object. For it was inferred from the fact that the entity at the receiving end of this relation has attributes different from those a physical object possesses that this entity cannot be identical with that physical object. If the so-called objects of perception enter into the picture only as members of the different possible states of affairs compatible with what one appears to be perceiving (one's immediate perceptions), no simple inference of this kind can be drawn.

This line of criticism is somewhat weakened by the fact that we have, in our ordinary usage, constructions with perceptual terms that do not prima facie fit into my view of these terms as expressing propositional attitudes. We have such locutions as '*x* perceives that *p*' or '*x* sees that *p*' where '*p*' is a placeholder for independent clauses, each of which specifies (is true in) a number of 'possible worlds'. This might be called the propositional construction or 'perceiving that' construction. But we also have locutions like '*x* sees *a*' where '*a*' is a free singular term, e.g., a proper name. The latter type of locution might be called the direct-object construction. It suggests a relationship between the perceiver and the objects of perception different from the one we have envisaged so far. The prevalence of this direct-object construction has probably also discouraged interest in the analogies between perceptual concepts and other modal notions. Part of what I have to do to defend my view of the logical behavior of perceptual terms is therefore to show that direct-object constructions with perceptual terms can be reduced to the 'perceiving that' construction.

V. AN ARGUMENT FROM INCOMPLETE PERCEPTUAL IDENTIFICATION

More interesting than any criticism of the argument from illusion is perhaps the observation that a different but closely related argument can be put forward for sense-data. This new argument is from our point of view considerably more intriguing than the original argument, although in its simple forms it is much less persuasive than the usual forms of the argument from illusion. This new argument might be called the 'argument from incomplete perceptual identification' rather than the argument from illusion. The situations it applies to are in fact much more commonplace than those considered in the argument from illusion. Consider, for instance, the following situation: There is a piece of chalk (say, *c*) on the table in front of someone (*d*); *d* perceives that it is white. We might express this as follows:

(8) d sees that $W(c)$.

Now suppose that c is in fact the smallest object on the table:

(9) $c = s$,

where 's' = 'the smallest object on the table'; and suppose further that d does not see that c is the smallest object on the table and that he does not see in any other way that the smallest object on the table is white:

(10) not: d sees that $W(s)$.

It is obvious that situations of this kind are perfectly possible, and in fact quite frequent. It may be argued that the possibility of such situations shows that what we are talking about are not ordinary physical objects like pieces of chalk. For if we were talking about them only, surely the trivial identity of c and s as physical objects ought to guarantee that exactly the same things can be said of them, i.e., that the terms 'c' and 's' are interchangeable everywhere (*salva veritate*). But this is just what is not the case in the situation we envisaged, for there the substitution of 's' for 'c' in (8) turns it into the statement

(11) d sees that $W(s)$,

which contradicts (10). Hence the objects of perception we are talking about here must be something different from ordinary physical objects. They might, for all that I can see, be labeled sense-data.

A closely related argument might run as follows: If what we are talking about in (8) through (10) were ordinary physical objects, we ought to be able to generalize existentially with respect to c in (8) and obtain

(12) (Ex) (d sees that $W(x)$),

where the bound variable x ranges over ordinary physical objects. But it can scarcely do so in (12), for what is the physical object whose existence makes (12) true? It cannot be c, for as a physical object c is identical with s, of which it is not true at all that d sees that it is white. But if it is not c, it is hard to see what this physical object could be. Hence the values of the bound variable x in (12) have to be something different from ordinary physical objects. There does not seem to be anything wrong with calling them sense-data.

VI. SENSE-DATA AS INTENSIONAL ENTITIES

It is obvious that this 'argument from incomplete perceptual identification' for the existence of sense-data presents us with a situation that is precisely analogous to the predicament into which the breakdown of the substitutivity of identity and of existential generalization put us in ordinary modal logic. The line of thought of those who were willing to posit sense-data on the basis of our argument would be analogous to the line of thought of those philosophers of logic who have been led to resort to intensional entities in understanding quantification into modal contexts. In other respects, too, there seems to be a great deal in common to the theory of perception and to the philosophy of modal logic. For instance, some philosophers apparently want to avoid speaking of 'what seems to be the case to someone' as a primitive or irreducible idea. They would like to see these locutions reduced to expressions in which we only speak of what is. These philosophers are from a logician's point of view so many unwitting allies of those modal logicians who would like to avoid all talk of possible worlds different from the actual one. Whoever suggests that to talk of how things look to us is to talk, not of ourselves, but of certain aspects of these (ordinary physical) things, of their looks,[18] is tacitly sympathizing with those philosophers of modal logic who are willing to countenance only referentially transparent senses of modal notions as being fully legitimate. In short, in the last analysis John Langshaw Austin may have been the Willard Van Quine of perception theory.

In spite of these similarities, the connection between intensional entities and sense-data may still seem somewhat tenuous. It is true that sense-datum theories have taken forms whose originators would disown all connection of their ideas with an argument from incomplete perceptual identification. Nevertheless, there are more similarities between sense-data and intensional entities than we have so far discovered. Pointing out some of them may perhaps reduce an impression of tenuousness.

For instance, it may be asked: Should not sense-data be *data* and not individuals (in the logical sense of the word)? Are we not misrepresenting their status by turning them from the facts perceived (apparently perceived) into the ultimate objects (individuals) to which perceived attributes belong?

One answer is that our conception of sense-data (if any) is essentially that of G. E. Moore and of a number of other prominent philosophers, however it may be related to that of less careful and explicit sense-datum theorists. In order to show this, it suffices to quote the way in which Moore introduces sense-data in his lectures on *Some Main Problems of Philosophy* in 1910–11. Moore is considering a certain visual impression of his. "These things: this patch of whitish colour, and its size and shape I did actually see. And I propose to call these things, the colour and size and shape, *sense-data*, things *given* or presented by the senses – given, in this case, by my sense of sight" (Moore's italics).[19] "These things", although intended to be *things*, are perhaps not yet individuals in the logical sense of the word. But Moore soon saw the light. When the lectures were published in 1953, Moore added the following remark to the quoted passage: "I should now make, and have for many years made, a sharp distinction between what I have called the patch, on the one hand, and the colour, size, and shape, *of* which it is, on the other; and should call, and have called, *only* the patch, *not* its colour, size, or shape, a 'sense-datum'." I cannot think of a clearer statement showing that sense-data were for Moore *individuals.*[20]

It may also be questioned whether the sense-data that someone might be inclined to introduce by an argument from incomplete perceptual identification can serve any of the epistemological purposes which sense-data are traditionally taken to serve (and sometimes specifically introduced to serve). In answering this query, I am somewhat handicapped by the fact that I do not believe that there are sense-data in any usual sense of the word, and hence cannot say what they perhaps might be good for. It is a fact, however, that if the characteristic features of intensional entities, as they are often conceived of by philosophers, are attributed to sense-data, we find ourselves ascribing to sense-data some of precisely those features that allegedly made sense-data so attractive epistemologically. For instance, Quine says (or used to say) that any two ways of characterizing one and the same intensional entity (in ordinary modal contexts) must be analytically (necessarily) equivalent.[21] '$i=j$' implies 'necessarily ($i=j$)' if i and j are intensional entities. The analogue to this would be to say that whenever two sense-data are in fact identical, they are perceived to be identical. It is, in this sense, impossible to make perceptual mistakes about the identity of sense-data. If there were such entities, they would be epistemologically privileged, at least in this sense. It might thus be said that sense-data are at least as respectable, and as difficult to avoid, as intensional entities are in modal logic.

We have already seen, however, that these are a pretty disrespectable bunch of entities. We have likewise seen that the kinds of argument which I labeled 'argument from incomplete perceptual identification' do not suffice to justify the conclusions they purport to justify. The features of logical behavior of perceptual concepts which apparently needed the postulation of sense-data can be accounted for by observing the character of these concepts as involving simultaneous reference to several possible states of affairs, along the lines indicated above for modal notions in general. For instance, we can generalize existentially with respect to the term '*s*' in the example above only if this term refers to one and the same individual in all the possible situations we have to consider here. Since these are all the possible states of affairs compatible with what *d* perceives, existential generalization is possible if and only if there is some individual *x* to which '*s*' refers in all these states of affairs. This means, however, that *d* sees that *s* is this individual *x*. Hence the extra premise needed is again of the form

(13) (Ex) (d sees that $(s = x)$).

Thus there does not seem to be any force in the arguments for sense-data we have considered so far. What has been said is not the whole story, however, and what remains to be said puts the matter into a somewhat different perspective.

VII. INDIVIDUATION AS A PREREQUISITE OF THE USE OF QUANTIFIERS

In the account I have given of the logic of propositional attitudes, I have so far disregarded certain very important presuppositions. I have said, for instance, that a free singular term is amenable to existential generalization if and only if it refers to one and the same individual in all the different 'possible worlds' we have to consider in the relevant context. Now this clearly presupposes that it makes sense to say that a member of one of the possible worlds *is the same individual* as a member of another possible world. In short, the account I have given of the logic of propositional attitudes and other modal notions presupposes that we can make what might be called cross-identifications, that is to say, identifications across the boundaries of possible worlds, or identifications between members of different possible worlds.

Since it was the identity of the respective references of a singular term in the different possible worlds we are considering that made it possible to say that it specifies a unique *individual*, the method of cross-identification which is presupposed in my account of the logic of propositional attitudes might also be called a method of *individuation* in contexts governed by propositional attitudes. Since variables bound to quantifiers range over individuals, a method of individuation is an indispensable prerequisite of all quantification into modal contexts. A quantifier that binds (from the outside) a variable occurring in a modal context does not make any sense without such a method of individuation, and its meaning is relative to this method.

At this point, it would be tempting to say simply that since quantification into modal contexts often makes perfectly good sense (even when these contexts are *not* construed transparently), we obviously must have as a part of our normal conceptual structure such method of individuation. That this is in fact the case is clear enough. But it is nevertheless worth one's while to take a somewhat closer look at the situation.

Consider, for instance, the concept of knowledge. Here the possible worlds we have to heed are described by all the different 'complete novels' compatible with what someone knows. It is clear that in most cases a comparison between two such novels will show fairly soon whether an individual figuring in one is identical with an individual described in the other. It is also pretty obvious what sorts of clues we would use in deciding this. They would be essentially the same kinds of leads we in fact use in reidentifying (in Strawson's sense) individuals.[22] What these are in the case of the different kinds of individuals is a difficult philosophical problem. (What constitutes personal identity, for instance?) It is amply clear, however, that for a wide variety of circumstances we have methods of cross-identification or individuation in the required sense, however difficult they are to describe with full philosophical clarity. Furthermore, it is clear that an attempt to describe these criteria is not the business of a poor modal logician. To describe the criteria of personal identity is part of the business of a philosopher of psychology, and the task of describing the other kinds of the individuation methods belongs to the province of other branches of philosophy, maybe to the philosophy of biology and of physics. I do not see much reason to worry whether suitable methods of individuation exist, although my philosophical colleagues may find plenty to worry about in the question of exactly how the methods we ordinarily rely on are to be described.

Essentially the same can be said of all the other propositional attitudes. The methods we use to individuate the objects of such attitudes are again essentially like our ordinary methods of reidentifying individuals, and hence relatively unproblematic for a logician.

Even a logician has to observe, however, that these methods of individuation rely heavily on certain contingent (non-conceptual) features of our environment. Without going into any detail, we still can see it is obvious that these methods of individuation turn on such facts as bodily continuity, continuity of memory, certain obvious features of the behavior of material bodies vis-à-vis space and time (one and the same body cannot be at two places at the same time; it takes time for it to get from one place to another; it does not change its shape or size instantaneously, etc.), and many similar physical and psychological regularities. To have a word for these methods of individuation, I shall call them *physical* methods of individuation or cross-identification. There may be good conceptual reasons why the methods of individuation which we ordinarily rely on make use of these regularities, but there does not seem to be any conceptual necessity that they, and only they, should be exploited for the purpose of individuating the objects of the various propositional attitudes. If certain doctrines of reincarnation were taken seriously, and if it really were possible to find out about people's earlier incarnations, our methods of individuation would have to be changed. If the equally improbable motto of the Wykehamists were literally true and manners were what "makyth man", other criteria of individuation than such mundane things as bodily continuity would be needed. Certainly we can imagine a primitive tribe performing a kind of Wittgensteinian language game in which the successive kings and medicine men are really one and the same person, irrespective of differences in looks and memories, as the successive Dalai Lamas are believed to be one and the same person by true believers.

VIII. PERCEPTUAL VS. PHYSICAL METHODS OF INDIVIDUATION

The possible multiplicity of methods of individuation is not of great interest, however, so long as it is not exhibited by our own conceptual system. Now the great interest of perceptual concepts for a philosopher of logic is due precisely to the fact that in connection with them we all as a matter of fact use two different methods of individuation. One of them is the method of physical individuation indicated above, but the other is essentially different from it. It seems to me that a great deal of the logic of perception is connected with this very fact.

What, then, is this other method of individuation? In order to see what it is, let us consider what someone, *d*, sees at some particular moment of time. Let us assume that he sees a man in front of him but that he does not see who the man is. Here the relevant 'possible worlds' are all the different states of affairs at the time in question that are compatible with everything he sees. We have already seen what it means to cross-identify individuals in different possible states of affairs by means of physical methods of individuation. By these methods, the man in front of *d* (let us call him *m*) is a different individual (different person) in some of the relevant possible states of affairs: just because *d* does not see who *m* is, the individual to whom the term '*m*' refers will be a different physico-psychological individual (different person) in some of the different states of affairs compatible with everything *d* sees then and there. In all these different states of affairs, however, there has to be a man in front of *d*. (Otherwise the state of affairs in question would not be compatible with what *d* sees.) The common perceptual relation of these different men to *d* separates them from the other individuals in each of the possible situations we are considering. Because of this, we may say that from the point of view of *d*'s perceptual situation they are after all one and the same man – the man in front of him.

This obviously can be generalized. When presented with descriptions of two different states of affairs compatible with what *d* sees, and with two individuals figuring in these two respective descriptions, we can ask whether they are identical as far as *d*'s visual impressions are concerned, and often we can answer this question. This question therefore gives us another method of individuating objects in contexts in which we are talking of what someone sees at a given moment of time, and a generalization to other perceptual terms is forthcoming.

We shall call individuals so cross-identified 'perceptually individuated' objects. Earlier we encountered 'physically individuated' objects. It would be suggestive and in many respects illuminating to call these two 'physical objects' and 'perceptual objects', respectively. In a way, this is just what is involved. What enabled us to say that the man in front of *d* is a unique visually individuated individual might be expressed precisely by saying that from *d*'s point of view there is in fact such a visual object as the man in front of him.

Striking though this way of speaking is, it is highly misleading. There is no question here of any ontological difference between different kinds of entities. The individuals which exist in the different possible worlds that we have to consider are of the same kind ontologically as the individuals existing in the actual world. There is no distinction between free singular terms referring to physically individuated objects and those referring to perceptually individuated objects. The only difference lies in the distinction between the two methods of individuation. This is a matter of the relation of the different possible states of affairs to each other. It does not appear as long as we are merely considering the different states of affairs one by one; it becomes relevant only when an implicit or explicit comparison between different states of affairs is made.

IX. TWO KINDS OF QUANTIFIERS AND THEIR MEANING

We have already seen that quantification into a context governed by a perceptual term involves such a comparison. Hence the meaning of quantifiers that from the outside bind variables occurring inside perceptual constructions (e.g., within the scope of the expression 'sees that') will depend on the method of individuation employed. In other words, when quantifying into perceptual contexts we have to reckon with two different pairs of quantifiers with different meanings. The variables bound to them range over the same sort of individuals, but differently individuated. We shall reserve the symbols '(Ex)' and '(Ux)' for quantifiers relying on physical methods of individuation. As quantifiers turning on perceptual methods of individuation, we shall use '$(\exists x)$' and '$(\forall x)$', respectively. Here it would again be tempting to say that variables bound to (Ex) and (x) range over physical objects while those bound to $(\exists x)$ and $(\forall x)$ range over perceptual objects. Saying this might even be illuminating for certain limited purposes. However, it will obscure the fact that ours is not simply a case of many-sorted quantification but that the relation between the two pairs of quantifiers is subtler than that. In general, in the kind of situation with which we are dealing, it is not illuminating to speak of quantifiers as *ranging over* a class of individuals. The conceptual situation is too complicated to be adequately described by this locution.

The distinction between different kinds of quantifiers is of considerable interest. One place where the distinction is relevant is a statement of perceptual identification. Since these turn on the use of quantifiers, we now have to distinguish between two sorts of perceptual identification. One of them will be expressed by a statement like

(7) (Ex) (d perceives that $a = x$),

while the other will be expressed by statements of the form

(14) $(\exists x)$ (d perceives that $b = x$).

What does the distinction between (7) and (14) amount to intuitively? This can be seen most clearly by considering cases in which (7) and (14) are true but only contingently (non-trivially) true. Cases in point are obtained by making a = the man in front of d and b = Mr. Smith. Then (7) will say that there is some physically individuated person x (individual) with whom a is identical in all the states of affairs compatible with what d perceives. In other words, d perceives that the man in front of him is this particular person ('physical object') x. This, clearly, is tantamount to d's *perceiving who* the man in front of him is. More generally, an approximate translation of (7) into a more idiomatic mode of discourse will be

(7a) d perceives what (or who) a is.

This is parallel to the familiar 'knowing what' or 'knowing who' construction which we have already met.

What, then, about the other kind of identification, typified by (14)? There it is said that one of d's perceptually individuated objects (his 'perceptual objects') is perceptually identified by *d* with Mr. Smith. In other words, *d* can (so to speak) find a place for Mr. Smith among his perceptual objects; Mr. Smith is one of his perceptual objects; in short, he *perceives Mr. Smith.* More generally, the appropriate translation of (14) into 'ordinary language' will be something like

(14a) *d* perceives *b*.

The difference between (7) and (14) is thus the same as the difference between *d*'s seeing who the man in front of him is and *d*'s seeing Mr. Smith.

Several comments are in order here. First of all, it is encouraging to see one of the distinctions which we have arrived at on the basis of abstract logical and semantical considerations to be reflected faithfully by perfectly ordinary language. This suggests strongly that we are on the right track here.

In fact, the only reason I have not been more categorical about the relation of our statements (7) and (14) to the vernacular locutions (7a) and (14a) is that these vernacular expressions often involve various existence and success presuppositions. We have all the methods at our disposal for incorporating these presuppositions into our formal statements (7) and (14) by adding suitable supplementary clauses. I shall not investigate here when and how they are to be added.

I should point out, however, that a statement involving a direct-object construction, e.g. (14a), is in ordinary usage sometimes construed as a statement about the individual in question, not about *b* 'whoever he is or may be'. According to what was said earlier toward the end of Section III, the force of the vernacular direct-object construction (14a) is then more likely to be expressible by

(15) $(\exists x)$ ($x = b$ and *d* perceives that x exists)

than by (14).[23]

Secondly, the translatability of (14) as (14a) or as (15) shows that we have now found an analysis of the direct-object construction in terms of quantifiers and of 'perceiving that'. The direct-object construction is therefore not an irreducible way of using perceptual terms. We have seen, on the contrary, that in order to spell out its precise meaning and its difference from the 'perceiving what (who)' construction we have to analyze carefully the presuppositions of the use of quantifiers in perceptual contexts. The presence of the direct-object construction in our ordinary language therefore does not go to show that objects can, logically speaking, enter into perceptual situations otherwise than as members of the different possible states of affairs we are implicitly considering here. Rather, on my analysis, it reinforces my point that this is the *only* way in which they enter into the logic of perception, and that there is no way out of the propositional-attitude character of our perceptual concepts.[24]

In terms of the behavior of singular terms vis-à-vis the two sorts of quantifiers, we can now make certain secondary distinctions between different kinds of free singular terms. It was already said above that the difference is *not* due to a difference in the individuals they refer to. There are not any strange entities here to be referred to. There may be differences between different kinds of free singular terms, however, in that the way in which some terms refer to the individuals they in fact refer to depends more on the perceptual situation, while the way other terms refer turns more on physical criteria and on other features independent of the particular perceptual situation we are considering.[25] On the logical level, this difference is betrayed by the fact that former kinds of singular terms are more likely to make (14) true when substituted for '*b*' than the latter, which conversely makes (7) true more often than the former when substituted for '*a*'. For some particular substitution values of '*a*' and '*b*', (7) and (14) might even be analytically true (i.e., true for conceptual reasons). In such cases, we might be tempted to extend the distinction between perceptual (perceptually individuated) objects and physical (physically individuated) objects to the references of free singular terms. More appropriately, we might perhaps speak of physically presented and perceptually presented objects. This would again be misleading, however, for the difference is between different kinds of singular terms, and not at all between their references. Moreover, even the difference between the terms is in evidence only when these terms are allowed to mingle with quantifiers.

With these qualifications in mind, it might nevertheless be illuminating to describe the difference between (7) and (14) as follows: In the former, a perceptually presented object is identified with a physical individual, whereas in the latter a physically presented object is identified with a perceptual individual.

Of course, it is perfectly possible that, unlike the terms we chose to instantiate '*a*' and '*b*' in (7) and (14), the former of these terms might rely predominantly on physical and the latter predominantly on perceptual methods of presentation. The only thing that happens then is that (7) and (14) become trivially true in most cases, and therefore less useful for our illustrative purposes than the statements we have considered.

A partial comparison with Miss Anscombe's paper is perhaps in order here. Our remarks on (14), (14a), and (15) show that the two kinds of constructions discussed in Section III (one used in making statements about a definite individual and the other used in making statements about whoever happens to be referred to by a singular term) are found with either of the two kinds of quantifiers. If the former difference is used to explicate Miss Anscombe's distinction between the material and intentional objects of perception, we thus have to say that her distinction cuts across our distinction between quantifiers that rely on physical methods of cross-identification and quantifiers that rely on perceptual methods of cross-identification. I have some difficulty in understanding fully Miss Anscombe's fascinating paper, but even so it seems to be clear that this cannot be the whole story. In many places Miss Anscombe seems to assimilate the distinction between intentional and material objects of perception to some kind of a distinction between perceptual and physical objects, a distinction which presumably has to be explicated in terms of a difference between different kinds of quantifiers or different kinds of occurrences of bound variables.

There need not be anything wrong with Miss Anscombe's distinction nor with my attempted reconstruction of it, apart from understandable vagueness. Miss Anscombe concentrates on the direct-object construction with perceptual terms. In our terms, this means that quantifiers relying on perceptual methods of individuation are being used. In such circumstances, the difference between the two constructions examined in Section III becomes largely a distinction between the use of singular terms (including variables bound to perceptual quantifiers) *outside* contexts governed by a perceptual term and *inside* such contxts. Outside such contexts, they merely serve to refer to an ordinary actually existing individual (or range over such individuals). By contrast, inside such a construction a bound variable involves the kind of 'perceptually individuated individual' which seems to be closely related to what Miss Anscombe has in mind, and in speaking of intentional objects of perception a free singular term can enter into an identity statement together with such a bound variable and in this sense also involve perceptual individuation. The difference between the roles of b in (14) and (15) is a clear case in point. Thus it is not hard to see how someone can easily assimilate to each other the distinctions between different kinds of quantification and different kinds of individuals when discussing the objects of perception. This assimilation obstructs clarity in this area, however, and it is advisable to keep the different distinctions separate as sharply as possible.

X. SENSE-DATA AS HYPOSTATIZATIONS OF PERCEPTUAL METHODS OF INDIVIDUATION

Perhaps the most interesting perspective opened by our observations is the possibility of appreciating what seem to me to be the deeper motives of the sense-datum talk. I see no reason to retract my earlier suggestion that there are no such members of our world as sense-data. However, we can perhaps now see one way in which they can easily steal their way into one's thinking, and that they have a certain justification. The closest legitimate approximation to sense-data that I can find are the values of those quantifiers that rely on perceptual methods of individuation. If their values could be reified into perceptual objects, these objects would be the legitimate heirs of sense-data. It seems to me justified to think of sense-data as having come about in this very way. Viewed in this light, sense-datum talk represents a dramatization of certain important features in the logical behavior of quantifiers relying on perceptual methods of individuation. The dramatic fiction of this sense-datum talk is a hypostatization of the values of variables bound to such quantifiers into alleged entities different from the ordinary ones.

How fully our quantifiers '$(\exists x)$' and '$(\forall x)$' serve the purposes sense-data were designed to serve is perhaps seen from the fact that their use can be justified by an argument which is but a slight variation of the argument from incomplete perceptual identification. Suppose I see a number of people but that I do not see who they are. Because of this failure, I cannot speak of them as those fully individuated persons (physically individuated individuals) who they (unseen by me, so to speak) in fact are. If I nevertheless want to speak of them as individuals (in the logical sense of the word), I must use other methods of individuation. This, in a nutshell, is the reason why perceptual methods of individuation are needed; and, recalling the argument from incomplete perceptual identification, it scarcely seems too far-fetched to say that it is also the true gist of this argument.

Thus viewed, sense-datum theories have the merit of a vigorous attempt to call our attention to certain interesting features of our conceptual system. They constitute a splendid example of revisionary metaphysics, incidentally illustrating most of the weaknesses of metaphysics *sans* logic. They anticipate perfectly valid logical distinctions, but exaggerate them beyond recognition. It might appear plausible to say that wherever there are different methods of individuation, there are also individuals of essentially different kinds, but this simply is not the case.

Although I thus find myself denying that the logical distinction between the two kinds of quantifiers has any ontological significance, I have a suspicion that a little more is at stake here than a pure logician is interested in. I suspect the kind of logical distinction I have made is exactly what those philosophers have had dimly in mind who have put forward and tried to defend such ontological distinctions as, for example, those between sense-data and physical objects. At any rate, if my suspicions are justified, we have reached a nice extension of Quine's dictum. Quine said that to be is to be a value of a bound variable. I suspect that to be in ontologically different senses is but to be a value of different kinds of bound variables.

XI. THE ELEMENT OF TRUTH IN SENSE-DATUM THEORIES

So far, I have emphasized that there are no sense-data in any ontologically relevant sense of the term. No individuals in any possible world are likely to include any such entities; they exist neither here nor there as far as ordinary existence as an 'inhabitant' of a possible world is concerned.

The other side of the coin is that methods of cross-identification inevitably create an objectively delineated supply of ways of individuating an object or person (in the context of some given propositional attitude of some definite person). They can be envisaged as functions (or partial functions) which from each possible world under consideration pick out (at most) one individual, *the same in all these worlds.* They are thus correlated one-to-one with the 'genuine' individuals we can speak of in the relevant context, and are therefore perhaps the closest counterpart to our intuitive idea of individual that we can incorporate in an explicit semantical theory.

In the case of d's beliefs, each of these functions is correlated with some singular term 'a' such that

$$(Ex)\,[d \text{ believes that } (a=x) \text{ and } (a=x)]$$

is true.

In spelling out the semantics of quantification into contexts governed by words for propositional attitudes we have to quantify over these functions (different ways of specifying a unique individual). In the case of quantifiers relying on perceptual methods of individuation, they will share some of the characteristics of the alleged sense-data. If they are what sense-data were intended to be, then there exist such things as sense-data.

This seems to be too hasty a conclusion, however. It is true that by Quine's criterion they would seem to be part of our ontology, since we have to quantify over them. This impression is misleading, however, and in fact brings out a clear-cut and important failure of Quine's dictum, construed as a criterion of *ontological* commitment. The functions in question are not inhabitants of any possible world; they are not part of the furniture of our actual world or of any (other) possible world. Thus it would be extremely misleading to count them in in any census of one's ontology. What they represent is, rather, an objectively given supply of ways in which *we* can deal with more than one contingency (possible world). They are part of our conceptual repertoire or our *ideology* (in something like Quine's sense) rather than part of our *ontology*. In a sense, we are committed to their *existence*, in the sense of their objectivity, but not to including them among '*what there is*' in the actual world or in any other world.

After a somewhat tortuous discussion, we have thus found a sense in which something like sense-data do exist. Yet it is only fair to say that we have also found that all the usual sense-datum theories are clearly wrong. They involve the fallacious hypostatization mentioned above and which we can now describe as an attempt to roll together the different values of these functions (which pick out individuals from different possible worlds) and reify them into something like ordinary individuals (of which one could ask such questions as what their relation to material bodies is).

For propositional attitudes other than perceptions, there exist by the same token 'intensional objects', however different those function-like entities are from ordinary bona fide individuals. A glimpse of the relation of these intensional entities to some traditional distinctions can perhaps be obtained from a quick comparison with Frege's formulations. His idea of the *sense* (*Sinn*) of a singular term, Frege avers, contains more than the idea of *reference* (*Bedeutung*), because in it we also have to include *the way in which the reference is given to us* (*die Art des Gegebenseins*).[26] Our intensional entities (remember that they are functions, not individuals) can be said to include not the way their references are given to us, but *the way in which they are* (*or can be*) *individuated.* From our point of view, an intensional entity – if the term is at all apt here, which I am rather dubious about – is a particular way of individuating an object, of specifying a unique, well-defined individual. They are objectively determined to the extent the truth values of statements containing such locutions as 'knows who', 'perceives who', 'has an opinion concerning the identity of', 'perceives (plus a direct object)', etc. are objectively determined.

This comparison with Frege also illustrates our disagreement with him. Any old way of picking out some individual or other from each possible world can be said to be a way of giving us an individual, but only if the individual is the same one in all the relevant possible worlds can it be said to amount to a way of *individuating* an object (or person).

The full import of these brief remarks can only be spelled out by describing in greater detail the semantics of perceptional terms and of other propositional attitudes. For reasons of space, it cannot be done here. Nor can the similarities and contrasts between perception and other propositional attitudes be examined in any further detail.

Suffice it to say merely that we can provide a partial answer to a question that no doubt has bothered you ever since we started comparing sense-data with intensional entities. If the arguments for both of them are parallel, why is there so much more of a palpable temptation to postulate sense-data than to postulate any shadowy intensional entities?

The obvious answer is that in the case of many other propositional attitudes there is nothing corresponding to perceptual methods of individuation. Since our approximations toward sense-data turned on a contrast between these methods and the ordinary physical methods of cross-identification, in the case of other propositional attitudes we do not have the same temptation to assume nonphysical entities as in the case of perception.

Some other propositional attitudes nevertheless allow methods of individuation that turn on the personal situation (or past situations) of the person in question rather than on physical criteria of cross-identification. Although I cannot here discuss them as fully as they deserve, it seems to me that for these other propositional attitudes, too, personal methods of individuation go together with the ubiquitous direct-object construction. Memory and to some extent knowledge are cases in point.

REFERENCES

[1] Mere formalization of the logical behavior of perceptual terms as a branch of modal logic is not by itself very important or interesting. What makes it promising is the existence of a well-developed semantical theory of modal logic. This is due largely to Saul Kripke and Stig Kanger; see Stig Kanger, *Provability in Logic*, Stockholm Studies in Philosophy, vol. I, Stockholm 1957; Saul A. Kripke, 'Semantical Considerations on Modal Logic', *Acta Philosophica Fennica* **16** (1963) 83–94; Saul A. Kripke, 'Semantical Analysis of Modal Logic I', *Zeitschrift für mathematische Logik und Grundlagen der Mathematik* **9** (1963) 67–96; Saul A. Kripke, 'Semantical Analysis of Modal Logic II', in *The Theory of Models*, Proceedings of the 1963 International Symposium in Berkeley (eds. J. W. Addison, L. Henkin, and A. Tarski), Amsterdam 1966. Cf. also my papers 'Modality and Quantification', *Theoria* **27** (1961) 119–28, and 'The Modes of Modality', *Acta Philosophica Fennica* **16** (1963) 65–81. (The last two papers are reprinted in the present volume, pp. 57–70 and 71–86, respectively.)

[2] See Section II. The characteristic behavior which is explained there in semiformal terms is precisely what the semantical theory of modal logic mentioned in the preceding reference strives to systematize.

[3] Delivered at the University of California, Berkeley, in 1963, and published in *Analytical Philosophy* (ed. by R. J. Butler), Second Series; Oxford 1965, pp. 158–80.

[4] Gilbert Ryle, *The Concept of Mind*, London 1949, pp. 222–23.

[5] Another natural possibility would be to follow my commentator's terminology and speak of 'perceptual belief' and 'perceptual knowledge'. The former is then what my 'perceives that' locution is in the first place supposed to cover. I am not quite sure, however, that this terminology is free from misleading connotations.

[6] These problems have been emphasized most persuasively by W. V. Quine; see the relevant parts of his *From a Logical Point of View*, Cambridge, Mass., 1953; *Word and Object*, Cambridge, Mass., 1960; and *The Ways of Paradox*, New York 1966.

[7] Richard C. Jeffrey, *The Logic of Decision*, New York 1965, pp. 196–97.

[8] See 'Form and Content in Quantification Theory', *Acta Philosophica Fennica* **8** (1955) 7–55, and 'Modality and Quantification'.

[9] These paraphrases are not quite accurate, however. They omit the 'fine structure' among the different possible worlds, which is due to the fact that not all possible worlds are legitimate 'alternatives' to a given one. For the significance and uses of this alternativeness relation, see my works 'The Modes of Modality'; *Knowledge and Belief*, Ithaca, N.Y. 1962; and 'Quantifiers in Deontic Logic', *Societas Scientiarum Fennica, Commentationes Humanarum Litterarum* **23** (1957), no. 4. Incidentally, most of the peculiarities of my somewhat loose use of quotes have undoubtedly caught the reader's eye. For stylistic ease, I am among other things pretending that the letters '*a*' and '*b*' *are* (particular) names or other free singular terms instead of *doing duty for* (arbitrary) names; and likewise for propositional 'variables'. Furthermore, quotes are omitted from displayed sentences.

[10] See 'Are Logical Truths Tautologies?' and 'Kant Vindicated', in *Deskription, Analytizität und Existenz*, 3–4 Forschungsgespräch des internationalen Forschungszentrums Salzburg, ed. by Paul Weingartner (Pustet, München und Salzburg 1966), pp. 215–33 and 234–53, respectively; also 'Are Mathematical Truths Synthetic *A Priori*?', *Journal of Philosophy* **65** (1968) 640–651.

[11] In an explicit semantical treatment, this is shown by the fact that these are the only 'possible worlds' we have to quantify over.

[12] The importance of the breakdown of these two rules of inference has been aptly emphasized by Quine in the works referred to above.

[13] This results by replacing one of the two occurrences of 'Walter Scott' in 'George IV knows that Walter Scott = Walter Scott' by 'the author of *Waverly*' while the other occurrence remains intact. Such partial replacements may seem queer, but are in fact vital in many other, unproblematic contexts.

[14] Cf., e.g., W. V. Quine, 'Quantifiers and Propositional Attitudes', *Journal of Philosophy* **53** (1956) 177–87; reprinted in *The Ways of Paradox*.

[15] If this were not the case, i.e., if bound variables did not range over genuine individuals, expressions (5) through (7) could scarcely play the role I have assigned to them. For if they failed to do so, the truth of (5) through (7) could not guarantee the kind of uniqueness of reference which is needed if these expressions are to serve as the extra premises that are to safeguard quantification into the modal contexts in question.

[16] For surveys and discussions of this argument, see, e.g., Konrad Marc-Wogau, *Die Theorie von Sinnesdaten*, Uppsala Universitets Årsskrift, Uppsala 1945; A. J. Ayer, *The Foundations of Empirical Knowledge*, London 1940; J. L. Austin, *Sense and Sensibilia*, Oxford 1962; Roderick Firth, 'Austin and the Argument from Illusion', *Philosophical Review* **73** (1964) 372–82; A. J. Ayer, 'Has Austin Refuted the Sense-Datum Theory?', *Synthese* **17** (1967) 117–40.

[17] Thomas Reid, *Essays on the Intellectual Powers of Man* (ed. and abridged by A. D. Woozley), London 1941, p. 145.

[18] Cf. Austin, *Sense and Sensibilia*, p. 43: "I am not disclosing a fact about *myself*, but about petrol, when I say that petrol looks like water.... Is it not that ... looks and appearance provide us with *facts* on which a judgement may be based ... ?"

[19] *Some Main Problems of Philosophy*, London 1953, p. 44. For the importance of this point for the rest of sense-datum philosophy, see *ibid.*, p. 45, n. 6.

[20] The same fact emerges clearly from the pronouncements of several of the other well-known sense-datum theorists. See, e.g., H. H. Price, *Perception*, London 1932, p. 64: "For we are acquainted with *particular instances* of redness, roundness, hardness and

the like, and such *instances of* such *universals* are what one means by the term sense-data" (my italics), or Bertrand Russell, *Mysticism and Logic*, London 1918, p. 147: "When I speak of a 'sense-datum', I do not mean the whole of what is given in sense at one time. I mean rather such a part of the whole as might be singled out by attention: *particular* patches of colour, *particular* noises, and so on" (my italics).

[21] See, e.g., Quine, *From a Logical Point of View*, pp. 151–52.

[22] Here we have something more than a mere similarity, namely, an analogy. In the one case (reidentification) we are dealing with identifications between members of temporally different states of affairs, in the other (different possible worlds) we normally identify individuals occurring under different possible courses of events. It is clear that some considerations are common to the two cases, although there obviously are also dissimilarities.

[23] Our frequent preference of (15) to (14) as a translation of (14a) is brought out by the fact that (14a) is often thought of as being subject to the substitutivity of identity: if *d* perceives *b*, he perceives it under any name or description. When this is the case, (15) is a better translation than (14).

[24] Cf. Moore's formulation of a closely related point: "We should then have to say that expressions of the form 'I believe so-and-so', 'I conceive so-and-so', though they undoubtedly express *some* fact, do *not* express any relation between *me* on the other hand and an object of which the name is in the words we use to say *what* we believe or conceive" (*Some Main Problems of Philosophy*, p. 288).

[25] Demonstratives are typical instances of the former, proper names of the latter.

[26] Gottlob Frege, 'Sinn und Bedeutung', *Zeitschrift für Philosophische Kritik* N.S. **100** (1892) 25–50, esp. 26–27.

DEONTIC LOGIC AND ITS PHILOSOPHICAL MORALS

I

Anyone who practices or preaches philosophical analysis might do well to study some deontic logic.[1] That is anyway the thesis of this paper. I shall suggest that a semantical approach to deontic logic offers as clear-cut examples as any analyst is apt to find anywhere of several key operations of his *métier*. They include:

(1) The use of our intuitions for the purpose of obtaining criteria of truth and/or consistency. (These will then also yield as a by-product rules of logical proof.)

(2) The re-education of some of our intuitions in the light of the semantical insights thus obtained.

(3) The interpretation (which sometimes amounts to a partial re-interpretation) of traditional concepts and doctrines within the framework the analysis has produced.

(4) The development of methods of bringing out the truth of our intuitions in subtle and roundabout ways. Even when there is a true gist to the intimations of our 'logical sense', its counsels often have to be codified in indirect ways. The true gist may turn out to be due to the logical status, not of the statement that we *prima facie* are concerned with, but of some other, related statement.

(5) The discovery of intrinsic ambiguities in some of the concepts we use in ordinary discourse.

(6) The exposure of fallacies to which one is led by overlooking these ambiguities.

II

I shall proffer illustrations of all these six maneuvers. Concerning the first I must try to be relatively brief and confine myself to giving a quick motivation to the semi-semantical conceptual framework I shall be relying on in the rest of the paper.

Here, as in so many other applications of the semantical approach, expositional vividness is enhanced by speaking of 'possible worlds'. (I have indicated elsewhere how this weird-looking notion can be stripped of its Leibnitian and other metaphysical overtones and reduced to a notion of sober model theory.[2]) Using it, we can say that to know what norms obtain (say in a given possible world **M**) is to know which possible worlds are in accordance with the norms that obtain in **M**. Let us call these possible worlds *deontic alternatives* to **M**. Let us apply the same terminology also the (partial) descriptions of the worlds in question.

The set of the descriptions of all possible worlds considered on one and the same occasion (in deontic logic) will be called a *model system* (of deontic logic). If we identify a partial description of a possible world with what I have called a *model set*, the notion of model system can tentatively be defined, for the purposes of deontic logic, by the following conditions on the set Ω in question:

(C.$\in$) Each member μ of Ω is a model set.

(C.O*) If $Op \in \mu \in \Omega$, and if $\nu \in \Omega$ is a deontic alternative to μ, then $p \in \nu$.

(C.O)$_{\text{rest}}$ If $Op \in \nu \in \Omega$, and if ν is a deontic alternative to at least one $\mu \in \Omega$, then $p \in \nu$.

(C.OO*) If $Op \in \mu \in \Omega$, and if $\nu \in \Omega$ is a deontic alternative to μ, then $Op \in \nu$.

(C.P*) If $Pp \in \mu \in \Omega$, then $p \in \nu$ for at least one deontic alternative $\nu \in \Omega$ to μ.

(C.o*) If $Op \in \mu\ \Omega$, then $p \in \nu$ for at least one deontic alternative $\nu \in \Omega$ to μ.

Here 'O' is shorthand for the normative prefix 'it ought to be the case that', 'P' for 'it is permissible that', and $\in$ is to be read 'is a member of'.

These conditions are all we need in deontic logic to characterize a model system. Thus a model system is an arbitrary set of model sets, together with an arbitrary relation defined on it (called the alternativeness relation), which jointly satisfy the conditions listed above. When this definition is supplemented by a characterization of a model set, we have an explicit definition of a model system.

The satisfiability of a set of sentences can be defined as its imbeddability in a member of a model system. In other words, a set of sentences λ is satisfiable if and only if there is a model system Ω and a model set $\mu \in \Omega$ such that $\lambda \subseteq \mu$. A sentence p is valid (logically true) if and only if $\{\sim p\}$ is not satisfiable. As usual, we shall say that q is a logical consequence of p if and only if $(p \supset q)$ is valid, i.e. if and only if $(p \,\&\, \sim q)$ is not satisfiable.

III

The intuitive motivation of the conditions that define a model system is not hard to fathom. Let us call the world described by μ **M**. Then according to (C.P*) it is permissible that p only if in some deontic alternative to **M**, i.e. in some possible world in which the norms (obligations, duties) of **M** are satisfied, it is the case that p. In brief, something is permissible only if it is compatible with all the norms – which is obvious enough. Condition (C.O*) simply spells out the idea that deontic alternatives to **M** are worlds in which all norms obtaining in **M** are satisfied. Condition $(\text{C.O})_{\text{rest}}$ makes the much subtler point that in order for a possible world (say **N**, defined as the world described by ν) to satisfy all the normative requirements obtaining in **M**, not only must the 'old' obligations obtaining in **M** be fulfilled, but also whatever 'new' obligations have 'come about' when we move from **M** to **N**. The need of this requirement becomes patent in connection with (C.P*): permissibility of p in **M** presupposes more than that it could be the case that p while all the overt norms of **M** are satisfied, i.e. that p could be realized without violating any of the duties that actually obtain in **M**. Often, a permission can in fact be made use of only at the expense of new duties which of course have to be fulfilled in **N** if the truth of p in **N** is to guarantee its permissibility in **M**. (These duties are of course based on the system of norms that obtains in **M**, but they need not be themselves overt duties in **M**.)

Another way of motivating the same point is to say that a deontic alternative like **N** to **M** is intended to be, from the point of view of those norms which obtain in **M**, a kind of *deontically perfect world.* This requirement of perfection means not only that all old duties (i.e. duties obtaining in **M**) are fulfilled in **N**, but all relevant duties are satisfied there. And naturally the duties obtaining in **N** itself are highly relevant here.

Condition (C.OO*) says that the norms obtaining in **M** will continue to obtain in its alternatives. This is a somewhat less obvious requirement than the previous ones. If (C.OO*) is adopted, it implies – together with $(\text{C.O})_{\text{rest}}$ – (C.O*), which becomes dispensable. However, in view of the lesser certainty of (C.OO*) I have also listed (C.O*) as a separate condition. (C.o*) has the force of saying merely that each norm can be thought of as being realized in some world or other.

Further conditions do not seem to be forthcoming. For instance, permissions ('permissibilities') cannot in any case be 'moved over' from μ to one of its alternatives, say v, in the way (C.OO*) says that norms are 'movable', for obviously one can quash a permission by making use of another.

It is important to realize that, and why, we often have to consider more than one deontic alternative to a given world, say **M**. This is due to the fact that not all permissions that obtain in **M** can be made use of in the same world. The simplest case in point is a deontically neutral proposition p, i.e. one for which $Pp \in \mu$, $P{\sim}p \in \mu$. Since p and ${\sim}p$ cannot both be the case in the same possible world, (C.P*) forces us to consider more than one deontic alternative to μ.

In formulating the semi-semantical conditions (C.∈)–(C.o*) above we did not have to consider explicitly iterated deontic operators (deontic operators occurring within the scope of other deontic operators). Consequently, whatever reasons there may be for adopting these conditions, they are independent of the problem of interpreting iterated deontic operators. For this reason, it is all the more remarkable that the conditions we have already accepted in effect give us a way of interpreting sentences containing iterated deontic modalities, simply by applying to them the same principles that apply to non-iterated ones. The controversies and problems concerning iterated deontic operators can thus be by-passed completely. In my judgement they merely illustrate the futility of attacking interpretational problems by trying to formalize ordinary language directly, without first developing suitable semantical tools which would show precisely how the formalization is to be undertaken.

IV

The notion of a model set μ ('partial description of a logically possible world') may be defined by the following requirements:

(C.$\sim$) If $p \in \mu$, not $\sim p \in \mu$.
(C.&) If $(p \& p) \in \mu$, then $p \in \mu$ and $q \in \mu$.
(C.$\vee$) If $(p \vee q) \in \mu$, then $p \in \mu$ or $q \in \mu$ (or both).
(C.E) If $(Ex)p \in \mu$, then $p(a/x) \in \mu$ for some individual constant *a*.
(C.U) If $(Ux)p \in \mu$ and if the free singular term *b* occurs in the sentences of μ, then $p(b/x) \in \mu$.

I am using '*x*', '*y*', '*z*', ... as bound individual variables and '*a*', '*b*', '*c*',... as (if they were) free singular terms. Furthermore, $p(a/x)$ is the result of replacing '*x*' everywhere by '*a*' in *p*, and similarly for other terms. For simplicity, it has been assumed in all our conditions that all connectives other than $\sim$, &, and $\vee$ have been eliminated and that all negation-signs are driven as deep into our formulae as they go, i.e. so as to precede immediately an atomic expression. This can always be achieved by means of the law of double negation, de Morgan's laws, the interdefinability of the two quantifiers (with the help of $\sim$), and the similar intertranslatability of *O* and *P*.

This translatability of e.g. *P* as $\sim O \sim$ shows that we are dealing with a pretty weak sort of permission here: permission to do *p* means simply absence of a prohibition to do *p* (=obligation not to do *p*). This explains my preference of the term 'permissibility' over the usual word 'permission'. There certainly are stronger notions than our permissibility, notions which exhibit a different kind of logical behavior. For instance, it may be suggested that inalienable *rights* are characterized by the very transferability which we saw fail for permissibility: to make use of such a right does not render any other inalienable right any less a right. However, in this essay I shall not discuss such stronger notions.

It has been assumed that we are dealing with an interpreted first-order language without identity enlarged by the two deontic operators '*O*', '*P*' (which are assumed to form statements when prefixed to statements). If identity is introduced, a couple of extra conditions are needed. The interplay of deontic operators and quantifiers may be argued to occasion a number of modifications in our conditions as they have so far been formulated. These modifications are beyond the purview of the present paper, however. More generally, the quantificational aspects of our conditions will not be relied on in the sequel, for virtually all points we shall take up pertain to deontic propositional logic.

V

The semantical or semi-semantical conditions which we have just explained are so central that they merit a number of further comments. They are based on an idea which in some form or other occasionally crops up in traditional moral philosophy, albeit often in a rather confused form.

Perhaps the most important version of this idea (or a group of ideas) is the notion of a 'Kingdom of Ends' (*Reich der Zwecke*) which we find in Kant.[3] It is occasionally characterized by him as a 'mere ideal' ('freilich nur ein Ideal') which is not realized but which we nevertheless must be able to think of consistently. This state of affairs would be realized, according to Kant, if all the maxims based on the categorical imperative *were followed without exception* (*op. cit.*, p. 438; Kant's italics).

Since all moral maxims are for Kant based on the categorical imperative, we can thus simplify and generalize a little and say that for us the 'Kingdom of Ends' is the world such as it would be if all and sundry rational beings always honored all their obligations (duties). In this respect, a Kantian 'Kingdom of Ends' is like a deontic alternative to the actual world. These deontic alternatives are also 'deontically perfect worlds' of sorts: all obligations, both these that obtain in the actual world and those that would obtain in such an alternative possible world, are assumed to be fulfilled in each of them.

Notice also that for Kant the categorical imperative is obviously the principle of all maxims, both of those obtaining in the actual world and of those that are followed in a 'Kingdom of Ends'. Thus the requirement that all these maxims are followed presumably implies that in a Kantian 'deontically perfect world' both the 'old' and the 'new' obligations are fulfilled, just as our conditions (C.O*) and $(C.O)_{rest}$ require.

Generally speaking, the deontic alternatives to a given world are related to it rather in the same ways as a Kantian 'Kingdom of Ends' is related to the actual world. From the point of view of this given world, they are realizations of the (normative) ideals obtaining in it somewhat in the same way as a *Reich der Zwecke*. We can say of them the same as was said by Kant of a notion closely related to that of a Kingdom of Ends, namely of the notion of an intelligible world (*Verstandeswelt*): "The concept of an intelligible world is therefore only a *point of view* (*Stand-*

punkt) which the reason finds necessary to adopt *in order to think of itself as an active being*" (*um sich selbst als praktisch zu denken* ... *op. cit.*, p. 458, Kant's italics). We could perhaps bring Kant's formulation a little closer to ours if we changed the last few words to read: in order to be able to think of itself as acting according to its normative principles. Once Kant says that in morality a *possible* world of ends is considered as if it were the actual world of nature (*op. cit.*, p. 436, footnote).

In a sense the notion of a deontic alternative therefore is thus a somewhat watered-down and relativized variant of the Kantian notion of a 'Kingdom of Ends'. It is a much weaker notion because our concept does not contain any reference to a particular moral principle, be it the categorical imperative, universalizability, or what not, in the way Kant's notion does. It is a relativized notion, for it refers to the possible world which a deontic alternative is alternative to (from the point of view of which it is so to speak considered). Moreover, a deontic alternative to a given possible world is not a unique entity, contrary to the way in which Kant seems to have looked upon his 'Kingdom of Ends'. On the contrary, normally there are several deontic alternatives to a given possible world in a model system. This difference between us and Kant reflects partly the more important role which is played by the notion of permission in our thinking as compared with Kant. For it is precisely the multiplicity of permissions that typically leads us to consider more than one deontic alternative, as was pointed out above.

VI

By means of the methods we have developed we can discuss several conceptual problems connected with normative notions. As a preparation for these application, we must first make an important distinction. It may be approached by asking: What does the validity of a statement of the form $(p \supset q)$ mean? According to our definitions above it means that the statement $(p \,\&\, \sim q)$ is not satisfiable. The intuitive meaning of this fact can be expressed by saying that p cannot be realized without *ipso facto* realizing q, too. When we are thinking of or discussing normative matters, we often – unwittingly – slip into discussing something else. Without noticing it, we concentrate our attention, not on what can or cannot be realized, but rather on what can or cannot be realized *without violating any obligations*. In the case at hand, we are tacitly considering whether p can be realized without realizing q *while all norms are satisfied*. In other words, we are interested in whether $(p \,\&\, {\sim} q)$ can be true, not in any old possible world, but in a *deontically perfect* world. This means that we are considering, not the satisfiability of $(p \,\&\, {\sim} q)$, but the satisfiability of $P(p \,\&\, {\sim} q)$, in other words, not the validity of $(p \supset q)$, but the validity of $O(p \supset q)$. If (and only if) the former sentence is valid q is usually said to be *logically implied* by p (to be a *logical consequence* of p). If (and only if) the latter sentence $O(p \supset q)$ is valid, we shall say that q is *deontically implied* by p (is a *deontic consequence* of p).

The point of view which served to connect our concepts with Kant's also offers us illustrations of what the notion of deontic consequence amounts to. In a logical consequence, we are asking what the realization of p entails in any arbitrary possible world. In a deontic consequence, we are asking what the realization of p entails in a 'deontically perfect world' or, in Kantian terms, in a 'Kingdom of Ends'. This formulation suggests a general, albeit simewhat vague reason why our intuitions frequently pertain to relations of logical consequence in the realm of logical relations between norms (and between norms and facts). It is frequently much easier to be categorical about how things ought to be, i.e. how they would be in a 'deontically perfect world', than to figure out the complex duties one as a matter of fact has in the actual world. Hence one is likely to have firmer intuitions, too, about the former than about the latter.

This distinction between deontic and logical consequence has been overlooked by most students of deontic logic, although it seems to be implicit in certain concepts that have frequently been used in traditional moral philosophy. This neglect is all the more fatal as it often makes a crucial difference whether the intuitions we seem to possess about deontic concepts and about their interrelations are to be formulated as deontic or as logical implications. The former is the case far more often than logicians have realized. If, in such cases, the intuitions in question are nevertheless forced on the Procrustean bed of logical implications, fallacies are bound to arise.

VII

The literature of deontic logic offers instructive and amusing examples of such fallacies. For instance, the following plausible-looking principle has been put forward:

(7) If we are obliged to do A, then if our doing A implies that we ought to do B, we are obliged to do B.

This certainly looks like a 'quite plain truth' of logic, and it was taken to be one by A. N. Prior in the first edition of his *Formal Logic.*[4] He formulated it essentially as follows:

(8) $(Op \& (p \supset Oq)) \supset Oq.$

Whatever obviousness may seem to accrue to (7) belongs in fact to the validity of the corresponding deontic consequence, i.e. to the validity of

(9) $O((Op \& (p \supset Oq)) \supset Oq).$

This is shown by the fact that (8) is not valid on the assumptions we have made, i.e. that its negation

(10) $Op \ \& \ ({\sim} p \vee Oq) \ \& \ P{\sim} q$

is satisfiable. This satisfiability is shown by the model system which consists of the following two model sets:

(11) $\{Op, ({\sim} p \vee Oq), P{\sim} q, {\sim} p, (Op \& ({\sim} p \vee Oq) \& P{\sim} q)\}$
(12) $\{Op, p, {\sim} q\}$

of which the latter is assumed to be a deontic alternative to the former. That this set of sets of sentences satisfies the defining conditions of a model system can be verified by inspection. (For simplicity, it is assumed here that (C.&) is extended so as to apply also to conjunctions with more than two members.)

What is even more interesting, our model system brings out the reason why (8) is not valid. Model set (11), which may be viewed as representing the actual world, contains both the sentence *Op* and the sentence ${\sim}p$. Using the terms employed in (7), this means that we can escape our obligation to do B simply by failing to carry out the earlier obligation to do A. There is nothing logically impossible or even logically awkward about such a course of events: lamentable as it may be, many of our obligations in fact remain unfulfilled. This 'escape' can only be prevented by requiring that all our obligations be fulfilled. But to require this is in effect to understand (7) as expressing a deontic consequence rather than a logical one, i.e. as expressing the validity of (9) and not of (8).

That we really have a deontic consequence here can be demonstrated by showing that the negation of (9), i.e.

(13) $P(Op \& (\sim p \vee Oq) \& P \sim q)$,

cannot occur in any member μ of any model system Ω. This can be accomplished by making the counter-assumption that it does so occur and by reducing this assumption *ad absurdum*. The following argument serves to show this and incidentally serves to illustrate the way in which the defining conditions of a model set and model system can be used to establish validity:

(14) $P(Op \& (\sim p \vee Oq) \& P \sim q) \in \mu \in \Omega$ (counter-assumption).

(15) $Op \& (\sim p \vee Oq) \& P \sim q \in \nu \in \Omega$.

This follows from (14) in virtue of (C.P*) for at least one deontic alternative ν to μ in Ω. Furthermore we have from (15) by (C.&)

(16) $Op \in \nu$

(17) $(\sim p \vee Oq) \in \nu$

(18) $P \sim q \in \nu$, hence

(19) $p \in \nu$.

Here (19) follows from (16) in virtue of $(C.O)_{rest}$. This condition is applicable because ν is a deontic alternative to μ.

Because of (C. $\vee$), (17) implies that either

(20) $\sim p \in \nu$

or

(21) $Oq \in \nu$.

But (19) and (20) violate (C. $\sim$). Hence we can only have (21). But (21) and (18) likewise violate (C. $\sim$), thus reducing the counter-assumption (14) *ad absurdum*.

(If you are hesitant about applying (C. $\sim$) to statements containing deontic operators, you may continue the argument as follows:

(22) $\sim q \in \xi$

from (18) in virtue of (C.P*) for some deontic alternative $\xi \in \Omega$ to v. But then we must have

(23) $q \in \xi$

from (21) in virtue of (C.O*). Here (22) and (23) violate once again (C.$\sim$), this time as applied to simpler – possibly atomic – statements.)

The successful *reductio ad absurdum* of our counter-assumption (14) shows the desired validity of (9). In our approach to deontic logic, the situation is thus the one I claimed it to be: (7) has to be interpreted as expressing a deontic rather than a logical consequence. Nor is the possibility of showing this restricted in any way by the peculiarities of our approach. In Prior's old system, the impossibility of adopting (8) as a valid logical principle is shown by the unnatural consequences of an attempted adoption. Prior deduces from this assumption a version of the so-called first paradox of commitment. This is not the only awkward consequence, however, nor the most awkward one. By substituting $(p \& \sim p)$ for q in (8) and by noticing the disprovability of $O(p \& \sim p)$ in most systems in deontic logic, including Prior's, we can readily deduce from (8) the striking theorem

(24) $Op \supset p.$

This says that all obligations are in fact fulfilled, i.e. that we are dealing with a 'deontically perfect world', precisely as I argued that we must in the first place assume that we are doing.

In the second edition of *Formal Logic*, Prior has given up (8), prompted by syntactical considerations of the kind just mentioned. (He attributes them to A. R. Anderson instead of my 1957 paper 'Quantifiers in Deontic Logic' where they first appeared.) This does not seem to bring out the full generality of the problem, however.

Prior is not the only logician who has been seduced by the usual syntactical (axiomatic and deductive) methods into formulating perfectly valid relations of deontic consequence as relations of logical consequence, with all the absurd consequences resulting from such a course. It was a rather inauspicious beginning for deontic logic that the very first axiom of its very first attempted axiomatization embodied this mistake. This axiomatization was offered by Ernst Mally in his monograph *Grund-*

gesetze des Sollens: *Elemente der Logik des Willens*, Graz 1926. His first axiom was essentially the following:

(25) $((p \supset Oq) \& (q \supset h)) \supset (p \supset Oh)$.

This may be criticized along the same lines as (8) was criticized above. Again one can 'escape' the obligation that h simply by failing to carry out the duty expressed by Oq. Again the fallacy involved here is illustrated by the strange consequences of the adoption of (25) as a logically valid principle. Mally himself in effect deduced from his axioms the same principle (24) as was seen to follow from the adoption of (8). This consequence might seem counter-intuitive enough to overthrow any axiomatization of deontic logic. Unfortunately, Mally was not deterred by this strange consequence of his axioms. Apparently he found his axioms so obvious as to be above suspicion. Instead, he resorted to the hopeless expedient of trying to explain away the absurdity of (24) on interpretational grounds. No wonder the subject was not carried further for a while after Mally.

What makes (25) seductive is the fact that h is a *deontic* consequence of $(p \supset Oq)$, $(q \supset h)$, and p. This can be verified without any trouble along the same lines as the validity of (9) was shown above. Likewise, a counterexample is easily constructed to show that (25) is not valid on the basis of the assumptions we have made in the present paper.

Another example is provided by a principle which was put forward by K. Grelling in his article 'Zur Logik der Sollsätze', *Unity of Science Forum*, January 1939, pp. 44–47, and which *prima facie* is (in the words of Prior) "not without certain intuitive plausibility":

(26) If the doing of A and B jointly necessitates the doing of C, then if we do A and are obliged to do B, we are obliged to do C.

If this is interpreted as a logical consequence, it is a *non sequitur*, for the conclusion can be avoided simply by assuming that the obligation to do B is not fulfilled. This can be blocked by requiring that we are dealing with a 'deontically perfect world', i.e. with a relation of deontic consequence. In fact, although the straightforward formalization

(27) $((p \& q) \supset h) \supset ((p \& Oq) \supset Oh)$

of (26) is invalid, it can easily be shown that h is a deontic consequence of the antecedent of (27) together with $(p \& Oq)$.

Some of the awkward consequences of the adoption of (27) as a logically valid principle have been pointed out by Prior. To them one can add the deduction of (24) from (27) by substituting $(p \& \sim p)$ for h, $\sim p$ for p, and p for q.

Some of these fallacies may be avoided by strengthening material implications into stricter ones. (According to Prior, this has been suggested by G. E. Hughes as a way out of difficulties here.) This cannot be done in all cases, however. For instance, there seems to be no hope of repairing the 'Grelling paradox' (26) along these lines. Hence this suggestion cannot constitute satisfactory diagnosis of what is involved in them.

These examples illustrate forcefully the need to distinguish deontic implications from logical ones. Our discussion of the distinction, and of the consequences of not needing it, is also calculated to illustrate several aspects of an analyst's art. Perhaps the most important aspect exemplified here is the one listed as (4) in the beginning of this paper. However, a reader will also find examples of (5)–(6) and (2) in the preceding pages.

VIII

Another application of the concept of deontic consequence brings us closer to traditional discussions of moral philosophy. In such discussions, a considerable role has been played by what is known as the principle that 'ought implies can' or in the original German terms as the 'sollen-können' principle. (For a discussion of this subject, see e.g. G. H. von Wright, *Norm and Action*, Routledge and Kegan Paul, London 1963, pp. 108–116, 122–125.) As the name indicates, the question here is whether all oughts are intrinsically canny, that is, whether each obligation presupposes a possibility of fulfilling it. The discussions of this problem one can find in the literature can scarcely be said to have resulted in any kind of consensus. If we express the concept of possibility by the modal operator M (and the associated concept of necessity by the operator N), it lies close at hand to think of the problem as being concerned with the logical status of statements of form

(28) $Op \supset Mp.$

It is easily seen that on the assumptions so far made (together with certain unproblematic assumptions concerning the notions of necessity and possibility alone, unrelated to deontic notions) (28) is not valid. Of course it can be made valid by adopting some further principles concerning the interplay between the notions of possibility and necessity and deontic notions. The merits and demerits of such principles would require a longer discussion than can be undertaken here. It nevertheless seems safe to say that no obvious and uncontroversial principle is forthcoming on the level at which we are here moving to restore the validity of (28). In a context of a logical discussion, it therefore seems advisable not to try to salvage the 'ought implies can' principle by means of additional assumptions.

It is perhaps worth emphasizing that a particularly forceful type of argument for some versions of the principle is inapplicable here. It is often said that 'ought implies can' because a man cannot be blamed for not doing what he cannot do. And if he cannot be blamed for not doing something, he cannot be under an obligation to do it. Hence his being under such an obligation presupposes that he can fulfill it.

Whatever the merits of this line of argument are, it is inapplicable here. What we are dealing with in the present paper are impersonal norms rather than duties or obligations that pertain to some particular person. (Whether, and if so how, the latter can be analyzed in terms of the former is a question which will not be taken up here.) But if so, the argument just sketched for the 'ought implies can' principle falls outside the scope of the present paper, too, for it trades essentially in obligations of some particular person. (Only by so doing can the crucial notion of blame be brought in. For no one in particular can be blamed for not fulfilling an 'impersonal' norm of the kind we are here dealing.)

Hence it may appear that very little can be said about the 'sollen-können' principle here. One simple point can nevertheless be made. Whatever the status of (28) is, there is no problem about the status of a closely related sentence. Even if it is the case that Mp is not a *logical* consequence of Op, it is without any doubt a *deontic* consequence of the latter. In other words, sentences of the form

(29) $$O(Op \supset Mp)$$

are valid already in virtue of the assumptions we have made, plus one unproblematic assumption concerning the logical behavior of the concepts of necessity and possibility.

This can be shown by showing that the negation of (29) is not satisfiable, i.e. cannot occur in a member μ of a model system Ω. A proof to this effect can be carried out reductively:

(30) $P(Op \& N{\sim}p) \in \mu \in \Omega.$

This is our counter-assumption which we hope to reduce *ad absurdum.* (In bringing the negation of (29) to the form here displayed, we have made use of the equivalence of ${\sim}Mp$ with $N{\sim}p$.)

(31) $(Op \& N{\sim}p) \in \nu.$

This follows from (30) in virtue of (C.P*) for some deontic alternative ν to μ in Ω.

(32) $Op \in \nu$ (from (31) in virtue of (C. &))

(33) $N{\sim}p \in \nu$ (from (31) in virtue of (C. &))

(34 $p \in \nu.$

This follows from (32) in virtue of $(\text{C.O})_{\text{rest}}$, which is applicable because ν is a deontic alternative to μ.

(35) ${\sim}p \in \nu.$

This follows from (33) by the scarcely disputable principle that whatever is necessarily true (in a given possible world) is true (there). But (34) and (35) contradict (C.$\sim$), completing the desired reduction, and thus establishing the validity of (29). This means establishing that *Mp* is a deontic consequence of *Op*. It is important to realize that the argument by means of which this was established does not in any way turn on assumptions concerning the interplay of deontic concepts with the notions of possibility and necessity.

Our result is in itself very simple, and may even appear trivial – after it has been established. (It ought to be the case that all duties are fulfilled. Hence it ought to be possible to fulfill them). Some additional interest is in any case lent to our observations by the possibility that the 'sollen-können' principle was perhaps right from the beginning intended, however dimly and inarticulately, as an expression of a deontic consequence rather than a logical consequence.

The principle was brought to prominence in moral philosophy by

Kant. Hence we have to ask: how did he conceive of it? Kant's explanations are not distinguished by their lucidity, but an unmistakable and recurrent turn of thought in Kant is in any case a connection between 'ought implies can' principle and the concept of *freedom.* (See e.g. *Critique of Pure Reason* A 807, *Critique of Practical Reason*, first edition, p. 54.) Moral freedom, for Kant, lies in the very fact that a man can act in the way he ought to act. On the other hand, Kant tells his readers that a man exercises this freedom in so far as he is a member of that noumenal world to which he occasionally assimilates his 'Kingdom of Ends' and which on any showing behaves like the latter. Thus the fact that the moral law is followed in that possible world which Kant calls the 'Kingdom of Ends' or the 'noumenal world' is for him a ground for claiming that it is possible for a man to follow the moral law. But if this is the case, Kant's principle obviously amounts to a deontic rather than to a logical consequence. What he is saying is not so much that an obligation logically implies a possibility to fulfill it, but rather that the necessity of being able to think (if only as an *Idee*) of all our obligations as being fulfilled in some one world (at least in the noumenal world or in the 'Kingdom of Ends') shows the possibility of human freedom and hence the possibility of acting in accordance with our duties. On the basis of our earlier remarks, these other possible worlds may be compared to our 'deontic alternatives', and the fulfillment of all relevant duties in them will correspond to what $(C.O)_{rest}$ (and in part also (C.O*)) required. It is hence no accident that this very condition $(C.O)_{rest}$ played an absolutely essential role in the above argument (30)–(35) by means of which we demonstrated the possibility of interpreting the 'sollen-können' principle as expressing a valid *deontic* consequence.

From this point of view, the obscurity of many of Kant's formulations will be but another illustration of the difficulty of telling deontic implications from logical ones – a difficulty from which modern philosophers have not been found exempt, either. Be this as it may, it seems to me unmistakable that the whole trend of Kant's thinking in moral matters strongly suggests interpreting his 'sollen-können' principle as a deontic rather than logical consequence.

IX

An interesting further problem is posed by the notion of commitment.

What is – or can be – meant by saying that a certain fact or act (let it be described by p) commits one to acting a certain way, described (say) by q?

The two obvious candidates for this role that can be expressed in our language are the following:

$$(36) \qquad O(p \supset q)$$

and

$$(37) \qquad p \supset Oq.$$

Much ink has been spilled in discussing the relative merits of these two explications. It has been spilled in vain, for the conclusion seems to me inescapable that our commonplace notion of commitment is intrinsically ambiguous between the two renderings (36) and (37) (plus, possibly, still others).

Our semantical insights enable us to appreciate the difference between (36) and (37). The former reconstruction in effect assimilates, in the special use where (36) is logically true, the notion of commitment to our earlier notion of deontic consequence. On this interpretation p commits us to q if it is impossible to realize p in a 'deontically perfect world' without realizing q, too. Since it has already been seen that our logical intuitions in the area of normative concepts often in effect pertain to relations of deontic consequence, it may be expected that also our ideas of commitment must often be spelled out in terms of (36) rather than in terms of its rival (37).[5]

Moreover, from this point of view we can see that the notorious paradoxes of (derived) commitment are but particular cases of the paradoxes of implication, and hence devoid of any special interest for a student of deontic logic. The paradoxes consist in pointing out that, if (36) is a satisfactory analysis of commitment, a forbidden act commits one to everything and that everything commits one to an obligatory act. In other words,

$$(38) \qquad O \sim p \supset O(p \supset q)$$

and

$$(39) \qquad Op \supset O(q \supset p)$$

are said to be valid – as they of course are in our approach. However, if the validity of (38) and (39) is looked upon from the point of view of our 'deontically perfect worlds', the appearance of a paradox is considerably diminished. In (38), it is true to say that *p* cannot be realized in a deontically perfect world without realizing *q* because *p* cannot be so realized *simpliciter*. In (39), *q* cannot be realized in a deontically perfect world without realizing *p*, for *p* has to be realized in any such perfect world in the first place. Thus the 'paradoxes' lose their sting against our interpretation (36), provided that we realize what precisely it contains. At worst we have a residual feeling of awkwardness which can be traced to the same sources as the usual 'paradoxes' of entailment (implication).

We might also look at the matter slightly differently. There is a little doubt that as long as *p* and *q* are normatively neutral (neither obligatory nor forbidden), (36) catches one sense in which we all frequently speak of commitment. One obvious reason why the notion of commitment is often employed is to prevent our actual world from departing from a deontically perfect world. If *p* is the case and if it commits us to *q* in the sense (36), then the actual world will not match the standards of deontic ideality unless *q* will also be the case. To avoid this is one major purpose which the announcement and enforcement of commitments of form (36) is calculated to serve.

Of course, when *p* and *q* are not neutral, this purpose may become otiose: if *p* is forbidden, a discrepancy between the actual world and deontically perfect worlds has opened as soon as *p* has been realized, irrespective of whether *q* is realized or not, and likewise for the case in which *q* is obligatory. When the notion of commitment is used in such unusual circumstances, it cannot usefully serve the purpose just indicated, if construed as in (36). If the notion is nevertheless seriously employed in such circumstances, some other purpose and hence some other interpretation must be presumed. Thus on our analysis the paradoxes of derived obligation (38)–(39) do not show that the interpretation (36) of commitment is misguided as an approximation of what is involved in our idea of commitment in many ordinary contexts. At most, they illustrate the fact that in those unusual circumstances with which (38)–(39) deal some notion of commitment different from (36) is tacitly presupposed.

X

This does not show, however, that (36) is always what people's informal verbal statements about commitments presuppose even in perfectly normal circumstances. In fact, there are good general reasons for thinking that often (36) is not the intended interpretation. For one thing, from (36) together with a purely factual statement no unconditional statements of obligation follows. For instance, p and $O(p \supset q)$ do *not* imply Oq. In this sense, commitments of the kind (36) do not admit 'detachment'. Yet on some occasions we certainly consider ourselves justified to carry out such a detachment and to announce, on the basis of a fact and a commitment, a definite non-conditional obligation.

A commitment for which this is possible must have something like the force of (37) rather than (36). Whenever an actual obligation follows from a commitment plus certain facts, some reconstruction along the lines of (37) rather than of (36) is thus presupposed. Such cases seem in fact to be quite common.

Objections have been made to (37) as an interpretation of the notion of commitment. For instance, it has been alleged that on this interpretation the realization of whatever is not in fact realized 'commits' one to everything, for

(40) $$\sim p \supset (p \supset Oq)$$

is valid.

The fact is that what creates the appearance of a paradox here is not so much the idea on which (37) is based as rather the desire to have some stronger implicational tie between p and Oq than a material implication in (37). It is certainly true that in many of people's everyday uses of the notion of commitment such a stronger tie is presupposed. However, it is not clear to what extent this presupposition is due to 'pragmatic' or 'conversational' implications rather than to the basic logical force of the expressions involved. In any case, the question concerning the nature of this stronger tie is independent of our study of deontic notions which is very well served – at least up to a point – by the simple material-implication explication (37).

XI

The differences between (36) and (37) – as well as the reasons for using both of them as alternative explications of the notion of commitment – are illustrated by conflicts of duty. Such a conflict may e.g. result from a promise. I give an honest promise (let this act be described by p) to bring it about that q (say, have a cup of coffee with you). Unknownst to me, however, my father has fallen ill, which creates an obligation to see him that overrules my earlier promise. It seems to me that moral philosophers have felt somewhat uncomfortable in discussing this kind of situation, and perhaps one can also see why. For the fact that the obligation created by my promise is overruled means that it is false to say *simpliciter* that Oq, i.e. that I am obliged to fulfill the promise. (By the same token, the commitment involved in my promise cannot be construed as (37)). Yet it is clear that not everything is morally all right if I have to break my promise, however firmly this particular course of action may be prescribed to me by the norms I abide by. I have somehow done something wrong. This 'moral failure' is what easily makes one hesitant to say that in such a case there is no absolute duty to keep the promise.

Our distinction between (36) and (37) enables us to see precisely what goes wrong in such a case. It is obviously and clearly true, even in the case of a promise overruled, that *in a deontically perfect world* such a promise cannot be given without keeping it. In such a world, p cannot be realized without bringing it about that q. Even if the act of promising does not give rise to an actual duty to keep the promise (e.g. because of other duties), it none the less remains true that in this sense giving a promise commits one to keeping it.

The sense of commitment involved here is clearly (36). Thus it may be said that we need sense (36) to account for the possibility that a perfectly genuine commitment (e.g. a valid promise) may be overruled by other obligations, while (37) is needed to do justice to the conceptual fact that it sometimes does result in *actual* duties.

XII

By this time the reader has – hopefully – begun to appreciate the difference between (36) and (37). At the same time, an especially attentive reader may also have had an experience of *déjà vu* – of recognizing something he recalls from the literature of moral philosophy.

In fact, I have already slipped a few times into a bit of conventional jargon by speaking in connection with (37) of actual or absolute obligations. To make uninhibited use of this jargon, the contrast between (36) and (37) is essentially that between *prima facie* duties (obligations) and *actual* (absolute or overall) duties or obligations.[6]

It is in fact obvious that the situations we considered for the purpose of illustrating the difference between (36) and (37) are of the same kind as those which the perpetrators of the traditional dichotomy have used as paradigm cases of the contrast between *prima facie* duty and actual duty. The main problem to which they have addressed themselves is likewise admirably accounted for by our distinction. This problem is the question as to how an obligation can be overruled and yet remain – in *some* perfectly good sense – a genuine obligation. Our answer to this question was already given. It is now seen to admit of a formulation in terms of the traditional distinction.

In order to obtain an explicit reconstruction of the distinction *prima facie* obligation vs. actual obligation, let us consider some set of normative principles whose conjunction is n. (The sentences formulating these principles may be of the form Oq, but they may also exemplify such more complex forms as (36) or (37).) Let us also assume that we have as our factual premises a set of descriptive statements whose conjunction is p. Then we shall say that on the basis of the set of norms n, q is a *prima facie* obligation if and only if

$$(41) \qquad O[(n \& p) \supset q]$$

is valid, i.e. if and only if q is a deontic consequence of $(n \& p)$. Likewise, there is (by definition) an actual obligation that q if and only if

$$(42) \qquad (n \& p) \supset Oq$$

is valid, i.e. if and only if Oq is a logical consequence of $(n \& q)$. Thus the distinction between *prima facie* obligations and actual obligations is closely related to the distinction between the notions of deontic consequence and logical consequence. The ambiguity of our intuitions *vis-à-vis* this distinction is probably the major reason why the notions of *prima facie* obligation and actual obligation have been distinguished so late and confused so often. (For a discussion of an example of such confusion, see the last few pages of the present essay.)

A logician is amused to find that an important philosophical distinction once again turns out to be based on an operator-switch, i.e. on the order of two different logical operations, in the case at hand, between O and $\supset$.

Sir David Ross, who more than anyone else has been instrumental in introducing the concept of *prima facie* duty (obligation) into contemporary moral philosophy, uses the term *absolute* duty instead of actual duty for its contrary. This is not incompatible with our reconstruction. The fact that in (41) the deontic operator O governs a conditional (if-then) sentence show in what sense *prima facie* obligations are in our view non-absolute or conditional.

If n does not contain any normative notions, there is a *prima facie* obligation that q if and only if q follows logically from the non-normative premise $(n \& p)$. This fact throws some light on the notion of a technical norm and on its relation to other kinds of norms.

XIII

A comparison with the usual explanations of the *prima facie*-actual distinction readily shows the close relationship of this traditional distinction to our reconstruction, although I am perfectly willing to admit that the traditional distinction has occasionally been put to uses which our reconstruction does not catch. If anything, it seems to me that some of the traditional moralists have been somewhat timid in following up the implications of the distinction. Even Sir David Ross, in giving examples of the failure of *prima facie* duties to give rise to actual (absolute) duties, does not emphasize strongly enough how often – and how easily – such *prima facie* duties as e.g. arise out of a promise can, *qua* actual duties, be overruled by other obligations. What makes moralists hesitant to say, in the case of a failure of this kind, that no *actual* duty obtains is undoubtedly the vague feeling that something goes morally wrong in such cases. We have already seen, however, that this feeling is sufficiently accounted for by pointing out the precise sense in which a *prima facie* duty is violated: something takes place that would not happen in a deontically perfect world. Indeed the *actual* breach of morality which takes place in such cases is typically different from a failure to satisfy a *prima facie* obligation. For instance, in the case of (say) promising the only conclusion we can detach (the only *actual* duty we can infer) is the actual duty not to give the kind of promise that will be overruled by other obligations, assuming that the duty to fulfill a promise is a *prima facie* one. In other words: although from p and $O(p \supset q)$ we cannot infer Oq, we can from p, $O(p \supset q)$, and $\sim Oq$ infer $O \sim p$.

Toward the end of this essay we shall meet one more instance of a philosopher's failure to see how easily *prima facie* obligations can obtain without any corresponding actual (absolute) obligation obtaining, and without anything going wrong with our logic.

A reason for the importance of *prima facie* obligations follows from our earlier remark that we are likely to have firmer views concerning how things ought to be, i.e. what a deontically perfect world looks like, than concerning the multiple interrelations of actual duties. For what *prima facie* obligations specify is precisely what happens in deontically perfect worlds.

A bonus we obtain as a by-product of our reconstruction of the distinction between *prima facie* duties and actual duties is a handy terminology for the distinction between the two kinds of commitment (36) and (37) which was discussed above at length. The former may be called – and from now on will be called – *prima facie commitments* and the latter *actual* or *absolute commitments*.

XIV

Armed with these observations and distinctions, which help to clarify the nature of commitment and the nature of *prima facie* duty, we can approach what seems to me the prettiest fallacy (or group of fallacies) one can find in recent philosophical discussion. This is the fallacy that underlies John Searle's famous attempt to show 'How to Derive Ought from Is'.[7]

Many of the details of Searle's subtle and suggestive paper are irrelevant to our concerns here. If we may simplify his main point a little, Searle claims that from a purely factual premise (an 'is') describing an act of promising together with the analytical (Searle uses the term 'tautological') premise that promises ought to be kept it follows that there is an obligation to keep the promise in question (an 'ought'). In short, an ought follows from an is plus a tautological and hence empty additional premise.

Let p be a statement to the effect that a certain particular promise is given and let q state that this particular promise is kept. We can relate Searle's discussion to our own earlier discussion by expressing his 'tautological' second premise by saying that giving the promise in question commits one to keeping it. Thus Searle's 'derivation of ought from is' is (on our oversimplified reconstruction) of the form

(43) $$\frac{\begin{array}{c} p \\ p \text{ commits one to } q. \end{array}}{Oq}$$

Searle emphasizes that a promise is (analytically) an act of placing oneself under an obligation to keep it. This undoubtedly brings out the analytical (tautological) character of the second premise of (43) especially clearly. However, it does not suffice to explain the precise logical form of the argument (43).

To begin with, we shall not worry about the alleged analyticity of the second premise of (43). The much more obvious trouble with (43) is its ambiguity, due to the ambiguity of the notion of commitment. According to what has been said earlier we have a choice between two readings of (43):

(43*) $$\frac{\begin{array}{c} p \\ O(p \supset q) \end{array}}{Oq}$$

and

(43**) $$\frac{\begin{array}{c} p \\ p \supset Oq \end{array}}{Oq}.$$

This distinction between two senses of (43) corresponds neatly to the two senses of Searle's locution 'placing oneself under an obligation'. In (43*) the obligation in question is a *prima facie* obligation, while in (43**) it is an absolute one.

XV

How is Searle's argument to be evaluated in view of this ambiguity? Earlier, it was hinted that perhaps the most common notion of commitment is something like (36). Accordingly, we might expect that the most plausible interpretation of Searle's argument is (43*). Unfortunately,

(43*) is not a valid inference. (This can be seen by conctructing a model system a member of which contains p, $O(\sim p \vee q)$ and $P\sim q$ – if the point is not obvious enough.)

In contrast, (43**) is a valid inference, indeed an instance of *modus ponens*. Does this show that Searle is right? No, it does not. This reconstruction is based on the assumption that the notion of obligation involved in one's obligation to keep one's promises is an absolute (actual) obligation. If we adopt this position, then it becomes dubious whether the second premise (37) of (43), interpreted as (43**), is really analytical, as Searle claims. At first blush, it certainly appears patently false to say that an act of promising entails (analytically!) an *absolute* (actual, overall) obligation to keep it. Saying this seems to overlook completely the possibility that the *prima facie* obligation which is admittedly created by the promise should be overruled by some perfectly valid competing obligation. It was precisely to account for this possibility that absolute (actual) obligations were distinguished from the *prima facie* ones in the first place. But if so, the second premise of (43**) is not analytical (and may in fact be contingently false).

Thus it seems that the second reconstrual (43**) of Searle's argument fails as badly as the first one to serve the purpose it was calculated to serve. Although it yields a formally correct piece of reasoning, the resulting second premise is not analytical. Hence the 'ought' conclusion does not follow from an 'is' alone, but only in connection with another (non-tautological) 'ought'.

However, this is not the only possible way of viewing (43**). One might try to insist, after all, that its second premise is analytical. Of course, this stratagem will succeed only if the first premise p can somehow or else be strengthened. This extra strength can be sought for in two different directions. We may either require more of the notion of promising than before, so much more indeed that (37) becomes analytical in the case at hand. Alternatively, we may want to conjoin p (the statement that a promise is given) with the statement (let us call it cp) that certain *ceteris paribus* conditions are satisfied. As far as the formal structure of the argument is concerned, the two suggestions result in parallel treatments, the second differing from the first only in that the role of p is now played by the conjunction $(p \,\&\, cp)$.

Let us examine the first line of thought first. In this case, the truth of *p* cannot any longer be ascertained simply by examining the person in question, his actions, and his circumstances. However clearly he says 'I promise', however sincere he is, and however fully all the other factually ascertainable presuppositions of successful promise-giving are satisfied, we still cannot say in the intended sense that he promised (the intended sense here being the one in which promising means shouldering an *absolute* obligation to keep it) unless it can also be established that there are no stronger competing obligations for him to fulfill. But to show this is not to establish a matter of fact. It means evaluating the overall normative situation so as to ascertain that a certain *prima facie* duty is not overruled. In brief, the truth of *p* no longer amounts to a matter of fact; *p* is no longer a factual premise but a normative one.

On this interpretation, there is no formal fallacy in (43**). Moreover, the second premise is analytical all right. However, now the first premise *p* is not a descriptive statement, but contains a normative component. Hence the argument again fails to provide us with a 'derivation of ought from is': now the first premise is no longer an 'is'.

This line of interpretation is somewhat unrealistic in any case, for no one is likely to maintain seriously as strong a notion of promising ('really' promising) as is required for it. No one is likely to maintain, that is, that it is part and parcel even of some unusually strong sense of promising that giving a promise means undertaking an *absolute* obligation to keep it. However, there is a much stronger temptation to attempt the other way out and replace *p* by $(p \& cp)$, that is to say, to maintain that although promising in itself does not entail an absolute duty to keep the promise, it does so provided that certain *ceteris paribus* conditions are satisfied. This does not make any difference, however, for then what was just said of *p* will apply *mutatis mutandis* to the conjunction $(p \& cp)$. For the reasons given, $(p \& cp)$ will not be a purely factual statement. If *p* does not contain any normative elements, then the *ceteris paribus* condition *cp* will be at least partly normative.

We might thus represent schematically the three interpretations of (43**) which we have considered as follows:

(44)
$$\begin{array}{ll} p & \text{(factual)} \\ p \supset Oq & \text{(non-analytical)} \\ \hline Oq & \end{array}$$

$$(44^*) \qquad \frac{\begin{array}{ll} p & \text{(normative)} \\ p \supset Oq & \text{(analytical)} \end{array}}{Oq}$$

$$(44^{**}) \qquad \frac{\begin{array}{ll} (p \,\&\, cp) & \text{(normative)} \\ (p \,\&\, cp) \supset Oq & \text{(analytical)} \end{array}}{Oq}$$

Although all these three represent logically valid inferences, they fail to provide us with a 'derivation of ought from is' in the intended sense.

The specious plausibility of assuming, in the third line of interpretation just mentioned, that the *ceteris paribus* condition *cp* can be taken to be factual is witnessed by Searle's adherence (essentially) to this line of defense. He formulates the second premise as follows: "All those who [promise, i.e.] place themselves under an obligation are, other things being equal, under an obligation." He explains the need of the qualifying clause here by saying that "we need the *ceteris paribus* clause to eliminate the possibility that something extraneous to the relation of 'obligation' to 'ought' might interfere." The interfering factors that Searle here labels 'extraneous' include competing stronger obligations, which the *ceteris paribus* clause must also eliminate in order to serve its purpose. But they cannot be ruled out by means of purely factual assumptions.

Searle nevertheless strives to maintain that "there is nothing necessarily evaluative about the *ceteris paribus* conditions". His argument hinges on the observations that "an evaluation [of the competing obligations] is not necessary in every case" and that "unless we have some reason to the contrary, the *ceteris paribus* condition is satisfied, and the question whether he ought to do it is settled by saying 'he promised'." This argument has no force, however. It is true that in some cases no intervention takes place, i.e. that in some normative situations there are no conflicting obligations. But to state that this is the case is to make a normative statement. Saying that *j* is deontically neutral $(\sim Oj \,\&\, \sim O \sim j)$ is as much a normative statement as saying that it is obligatory or forbidden. Likewise, to say that a *prima facie* obligation is not overruled by others (and that the question of actual duty can be decided in the way Searle says) is to make a normative statement, however negative. And to try to tie the need of evaluation to the question whether counter-arguments have in fact been presented, or whether we (actually?) have 'reasons to the contrary', as Searle appears to do, is of course beside the point. The question is what is implied by the norms one has accepted, not what arguments have actually been put forward or what reasons one actually has.

XVI

This by no means exhausts the interest of Searle's clever paper, nor even the different types of argument he is considering. A closer examination of these alternative arguments would uncover flaws in them similar to those we have already discussed.[8] However, our sole purpose here is to illustrate such important distinctions as (36)–(37) by means of Searle's main argument, which does not motivate a discussion of the further details of his paper.

One can nevertheless use our distinctions also to emphasize the extent to which Searle is perfectly right. If $O(p \supset q)$ is a principle of one's normative system, however analytical, then one can after all infer from p that there is a *prima facie* obligation to bring it about that q. (To see this, put $O(p \supset q)$ for n in (41) and try to assume that its negation is satisfiable.) Hence Searle is right in a rather striking sense. He has in effect pointed out that from an 'is' and from an analytical principle one may legitimately derive a perfectly genuine obligation, viz. a *prima facie* obligation. An 'ought' does follow from an 'is', albeit only a *prima facie* 'ought'. This observation becomes all the more important in the light of our earlier observation that such a *prima facie* 'ought' is often what our intuitions are all about anyway.

From this point of view, the basic flaw of Searle's paper does not consist so much in putting forward a fallacious argument as in failing to spell out the sense in which his (correct) conclusion is to be understood. He is calling our attention to a perfectly legitimate relation of *deontic* consequence but discussing it as if it were a logical consequence. His argument thus illustrates once again a type of confusion which we have already noted several times in the course of the present paper.[9]

This does not completely spoil Searle's main purpose, however. Part of his general emphasis I can in fact share whole-heartedly. This part is the subtlety and multiplicity of the ways in which normative and factual concepts are interrelated. What we have discovered in the present paper tends to reinforce rather than to lessen this impression of interrelatedness.

I believe that this impression is fully justified, and that it will be strengthened by further studies. (For instance, such studies may be expected to bring to light the large extent to which the meaning postulates of many salient concepts contain both factual and normative elements.)

At the same time Searle's paper fails to establish another part of his aim, provided that the results of our critical examination of Searle are justified. If we are right, his argument does not show that we cannot carry out a sharp distinction between 'ought' and 'is' in the sense that in an appropriate explanatory model of our normative discourse the non-descriptive element is compressed into the deontic operators *O* and *P* and that the logical laws governing these operators will obey important 'conservation principles' reminiscent of Hume's dictum on 'ought' and 'is'. In these respects, our discussion is more likely to comfort Hume than Searle. If these questions are to be emphasized, we shall have to say (I have suggested) that an 'ought' does not follow from an 'is'.

XVII

There still remains the question: How are all these observations supposed to illustrate the six tricks of an analyst's trade that were listed in the beginning of my paper? I hope that the reader's answer is the same as mine, which *in nuce* is the following:

The relation of our intuitive ideas to their semantical counterparts was illustrated in Section III. Whatever reformation of the reader's intuitions I may have succeeded in accomplishing has most likely taken place as a result of my comments on the so-called paradoxes of commitment in Section IX (and perhaps also in Section X). The semantical framework developed here was related to assorted traditional concepts and problems in Sections V ('*Reich der Zwecke*'), VIII ('ought implies can'), and XII (*prima facie* obligation *vs.* actual obligation).

The necessity of formulating apparent logical relationships in terms of relations of deontic consequence rather than relations of plain logical consequence was first mentioned in Section VI and illustrated repeatedly throughout the rest of the paper. The ambiguity between deontic and logical consequence was seen in Section VII to have caused several outright mistakes in earlier discussion, and the ambiguity which there seems to be in our commonplace notion of commitment was discussed in

Sections IX to XI. In Sections XIV to XVI, an attempt was made to pin this ambiguity on John Searle's famous – or notorious – 'derivation of ought from is'.

REFERENCES

[1] Some of the general philosophical and methodological issues that arise in connection with the application of logical and semantical methods to the analysis of concepts expressed in ordinary-language-terms are discussed in my paper, 'Epistemic Logic and the Methods of Philosophical Analysis', *Australasian Journal of Philosophy* **46** (1968) 37–51, reprinted as the introductory essay of the present volume.

[2] See essays 2, 4–7 of the present volume and 'Form and Content in Quantification Theory', *Acta Philosophica Fennica* **8** (1955) 11–55. A systematic treatment of first-order logic from this point of view has been given by Raymond M. Smullyan in *First-Order Logic, Ergebnisse der Mathematik und ihrer Grenzgebiete*, vol. XLIII, Springer-Verlag, Berlin, Heidelberg, and New York, 1968.

[3] Immanuel Kant, *Grundlegung zur Metaphysik der Sitten*. We refer to this work in terms of the pagination of the second edition of the original.

[4] Arthur N. Prior, *Formal Logic*, Clarendon Press, Oxford, 1957; 2nd ed., 1962.

[5] The main difference between the distinction (36)–(37) and the earlier distinction deontic consequence vs. logical consequence is of course that neither (36) nor (37) has to be true for logical (conceptual) reasons, whereas the latter distinction dealt with two kinds of logical (conceptual) connections between statements.

[6] The primary sources of this distinction in recent moral philosophy are the writings of Sir David Ross, especially *The Right and the Good*, Clarendon Press, Oxford, 1930, and *The Foundations of Ethics*, Clarendon Press, Oxford, 1939.

[7] John R. Searle, 'How to Derive Ought from Is', *Philosophical Review* **73** (1964) 43–58, reprinted in Jerry H. Gill (ed.), *Philosophy To-Day* no. 1, The Macmillan Co., New York, 1968, pp. 218–235, and in Philippa Foot (ed.), *Theories of Ethics, Oxford Readings in Philosophy*, Oxford University Press, Oxford, 1968, pp. 101–114.

[8] A case in point is the following: Searle says that the kind of criticism I just presented is in any case inconclusive, "for we can always rewrite the relevant steps ... so that they include the *ceteris paribus* clause as part of the conclusion".

This sounds fine. However, everything depends on the precise way in which the incorporation of the *ceteris paribus* condition in the conclusion is supposed to be accomplished. There are two possibilities which yield essentially the following putative arguments:

$$(*)\qquad \frac{\begin{array}{c} p \\ p \supset O(cp \supset p) \end{array}}{O(cp \supset p)}$$

$$(**)\qquad \frac{\begin{array}{c} p \\ p \supset (cp \supset Op) \end{array}}{(cp \supset Oq)}$$

Now the second premise of (∗) is obviously false. Surely it does not follow from the fact that a promise is given in the actual world that in all deontically perfect worlds cp is followed by q. Hence (∗) must be ruled out.

The only way to avoid this conclusion is to make $cp \supset p$ a logical (analytical) truth. This, however, deprives (*) of all relevance for Searle's purpose.

In (**), the conclusion is a conditional with tacitly normative antecedent and an explicitly normative consequent. That such a statement follows logically from a factual statement together with an analytically true premise has no implications whatsoever for the is-ought distinction, any more than (say) the logical truth of $Oq \supset Oq$ has.

[9] (Added in proof.) In his new book, *Speech Acts*, Cambridge University Press, Cambridge, 1969, p. 181, Searle now distinguishes between two kinds of obligations, exemplified by

> All things considered, Jones ought to pay Smith five dollars

and

> As regards his obligation to pay Smith five dollars, Jones ought to pay Smith five dollars,

respectively. He says that only obligations of the latter type, not of the former, can be derived from an 'is'. This distinction comes very close to our distinction between absolute obligations and *prima facie* ones. Searle fails to spell out, however, precisely what is involved in the latter. I claim that when this is done, the limitations of Searle's argument become patent. A derivation of a *prima facie* ought from an 'is' does not violate the fact-norm dichotomy, correctly understood.

NOTE ON THE ORIGIN OF THE DIFFERENT ESSAYS

Of the essays included in this volume, 'Epistemic Logic and the Methods of Philosophical Analysis' first appeared in *Australasian Journal of Philosophy* **46** (1968) 37–51. Only a couple of minor changes have been made here.

'Existential Presuppositions and Their Elimination' is very nearly identical with 'Studies in the Logic of Existence and Necessity. I: Existence', *The Monist* **50** (1966) 55–76. Only a few additions and a couple of minor changes have been made here. Its planned sequel became, after a few metamorphoses, the essay 'Existential Presuppositions and Uniqueness Presuppositions' which is reprinted in the present volume.

'On the Logic of the Ontological Argument' is my contribution to the volume *The Logical Way* (ed. by K. Lambert) commemorating Henry Leonard (Yale University Press, New Haven 1969).

As printed here, 'Modality and Quantification' is a considerably expanded version of its namesake in *Theoria* **27** (1961) 119–128.

'The Modes of Modality' is a virtually intact reprint from *Acta Philosophica Fennica* **16** (1963) 65–82.

'Semantics for Propositional Attitudes' has also appeared in *Philosophical Logic* (ed. by J. W. Davis, D. J. Hockney, and W. K. Wilson), D. Reidel Publishing Company, Dordrecht 1969, pp. 21–45. A greatly abbreviated and modified version appeared under the title 'Meaning as Multiple Reference' in the *Proceedings of the Fourteenth International Congress of Philosophy*, vol. I, Herder-Verlag, Vienna 1968.

'Existential Presuppositions and Uniqueness Presuppositions' was originally my contribution to the Irvine Colloquium in May 1968 and will appear in the proceedings of that meeting, to be edited by K. Lambert and published by D. Reidel Publishing Company of Dordrecht, Holland.

'On the Logic of Perception' has grown out of my unpublished contribution to the Fourth Scandinavian Congress of Philosophy. The present version was read at the 1967 Oberlin Symposium in Philosophy, and appeared in the proceedings of that meeting which were published in 1969 as *Perception and Personal Identity* by The Press of Case Western Reserve University, pp. 140–173. The editors were Norman S. Care and Robert M. Grimm.

'Deontic Logic and Its Philosophical Morals' is previously unpublished in its present form. It includes, however, a few paragraphs lifted from my old paper 'Quantifiers in Deontic Logic', *Societas Scientiarum Fennica, Commentationes humanarum litterarum* **23** (1957), no. 4.

All previously published material is reprinted here with the appropriate permission. For these permissions I am grateful to the Editor of *Australasian Journal of Philosophy*, to the Editor of *The Monist*, to Professor K. Lambert as well as to Yale University Press, to the Editors of *Theoria*, to Societas Philosophica Fennica, to Professors Davis, Hockney, and Wilson, to Professors Care and Grimm as well as to the Press of Case Western Reserve University, and to Societas Scientiarum Fennica.

INDEX OF NAMES

INDEX OF SUBJECTS

SYNTHESE LIBRARY

Monographs on Epistemology, Logic, Methodology,
Philosophy of Science, Sociology of Science and of Knowledge, and on the
Mathematical Methods of Social and Behavioral Sciences

‡D. DAVIDSON and J. HINTIKKA: (eds.), *Words and Objections: Essays on the Work of W. V. Quine*. 1969, VIII+366 pp. Dfl. 48.—

‡J. W. DAVIS, D. J. HOCKNEY, and W. K. WILSON (eds.), *Philosophical Logic*. 1969, VIII + 277 pp. Dfl. 45.—

‡ROBERT S. COHEN and MARX W. WARTOFSKY (eds.), *Boston Studies in the Philosophy of Science*. Volume V: *Proceedings of the Boston Colloquium for the Philosophy of Science 1966/1968*. 1969, VIII + 482 pp. Dfl. 58.—

‡ROBERT S. COHEN and MARX W. WARTOFSKY (eds.), *Boston Studies in the Philosophy of Science*. Volume IV: *Proceedings of the Boston Colloquium for the Philosophy of Science 1966/1968*. 1969, VIII + 537 pp. Dfl. 69.—

‡NICHOLAS RESCHER, *Topics in Philosophical Logic*. 1968, XIV + 347 pp. Dfl 62.—

‡GÜNTHER PATZIG, *Aristotle's Theory of the Syllogism. A Logical-Philological Study of Book A of the Prior Analytics*. 1968, XVII + 215 pp. Dfl. 45.—

‡C. D. BROAD, *Induction, Probability, and Causation. Selected Papers*. 1968, XI + 296 pp. Dfl. 48.—

‡ROBERT S. COHEN and MARX W. WARTOFSKY (eds.), *Boston Studies in the Philosophy of Science*. Volume III: *Proceedings of the Boston Colloquium for the Philosophy of Science 1964/1966*. 1967, XLIX + 489 pp. Dfl. 65.—

‡GUIDO KÜNG, *Ontology and the Logistic Analysis of Language. An Enquiry into the Contemporary Views on Universals*. 1967, XI + 210 pp. Dfl. 34.—

p.t.o.

*EVERT W. BETH and JEAN PIAGET, *Mathematical Epistemology and Psychology*. 1966, XXII + 326 pp. Dfl. 54.—

*EVERT W. BETH, *Mathematical Thought. An Introduction to the Philosophy of Mathematics*. 1965, XII + 208 pp. Dfl. 30.—

‡PAUL LORENZEN, *Formal Logic*. 1965, VIII + 123 pp. Dfl. 18.75

‡GEORGES GURVITCH, *The Spectrum of Social Time*. 1964, XXVI + 152 pp. Dfl. 20.—

‡A. A. ZINOV'EV, *Philosophical Problems of Many-Valued Logic*. 1963, XIV + 155 pp. Dfl. 23.—

‡MARX W. WARTOFSKY (ed.), *Boston Studies in the Philosophy of Science*. Volume I: *Proceedings of the Boston Colloquium for the Philosophy of Science, 1961–1962*. 1963, VIII + 212 pp. Dfl. 22.50

‡B. H. KAZEMIER and D. VUYSJE (eds.), *Logic and Language. Studies dedicated to Professor Rudolf Carnap on the Occasion of his Seventieth Birthday*. 1962, VI + 246 pp. Dfl. 24.50

*EVERT W. BETH, *Formal Methods. An Introduction to Symbolic Logic and to the Study of Effective Operations in Arithmetic and Logic*. 1962, XIV + 170 pp. Dfl. 23.50

*HANS FREUDENTHAL (ed.), *The Concept and the Role of the Model in Mathematics and Natural and Social Sciences. Proceedings of a Colloquium held at Utrecht, The Netherlands, January 1960*. 1961, VI + 194 pp. Dfl. 21.—

‡P. L. R. GUIRAUD, *Problèmes et méthodes de la statistique linguistique*. 1960, VI + 146 pp. Dfl. 15.75

*. J. M. BOCHEŃSKI, *A Precis of Mathematical Logic*. 1959, X + 100 pp. Dfl. 15.75

Sole Distributors in the U.S.A. and Canada:

*GORDON & BREACH, INC., 150 Fifth Avenue, New York, N.Y. 10011
‡HUMANITIES PRESS, INC., 303 Park Avenue South, New York, N.Y. 10010